SEX, AN EXPLORATION OF SEXUALITY, EROS AND LOVE

ISBN: 978-1-291- 56931-5

Andreas Sofroniou   2013 ©   Copyright

All rights reserved.
Making unauthorised copies is prohibited.
No parts of this publication may be reproduced, transmitted, transcribed, stored in a retrieval system, translated in any language, or computer language, in any form, or by any means, without the prior written permission of Andreas Sofroniou.

Andreas Sofroniou   2013 ©   Copyright

SEX, AN EXPLORATION OF SEXUALITY, EROS AND LOVE

ISBN: 978-1-291- 56931-5

CONTENTS

	PAGE:
PREFACE	*4*
1. MYTHOLOGICAL EROS	*5*
EROS, CUPID AND SEX	*5*
SYMBOLISM OF SEX AND THE LIFE CYCLE	*7*
SEX, SEXUALITY AND REPRODUCTION	*7*
LEGAL REGULATION	*7*
SEXUAL BEHAVIOUR	*9*
FOLK ART AND VARIATIONS BY SEX	*10*
ECOLOGICAL INFLUENCES	*10*
2. BIOLOGICAL AND PHYSIOLOGICAL ASPECTS OF SEX	*11*
SEXUAL AND NON-SEXUAL REPRODUCTION	*11*
ASEXUAL REPRODUCTION	*12*
SEX DETERMINATION	*12*
SEX DIFFERENCES	*13*
SEX CELLS	*13*
SEX CHROMOSOMES	*14*
ABNORMAL CHROMOSOME EFFECTS	*16*
PSEUDOHERMAPHRODITISM	*16*
TRANSSEXUALISM	*17*
SEX DIFFERENCES IN ANIMALS	*18*
SYSTEM, IN FROG, MOUSE, AND MAN ALIKE	*18*
SEX DETERMINATION	*19*
SEX PATTERNS	*20*
SEX HORMONES	*21*
PRODUCTION	*21*
OESTROGEN	*22*
FERTILIZATION, SEX DETERMINATION, AND DIFFERENTIATION	*24*
HISTORY OF SEX HORMONES	*25*
CHEMICAL COMPOUND OF SEX HORMONES	*25*
PERCEPTION OF SEX	*26*
ADAPTIVE SIGNIFICANCE OF SEX	*27*
FRIGIDITY	*28*
HEREDITY AND SEX LINKAGE	*29*
HORMONES IN ANIMALHOOD	*31*
SEASONAL OR PERIODIC SEXUAL CYCLES	*33*
SEX AND THE EFFECTS OF ENVIRONMENT	*35*
HUMAN DEVELOPMENT AND SEX DIMORPHISM	*35*
DIFFERENTIATION OF THE SEXES	*36*
HUMAN EMBRYOLOGY AND THE GENITAL SYSTEM	*38*
GONADS	*38*
SEXUALITY: COMPLEMENTARY MATING TYPES	*38*
DIRECT ANIMAL DEVELOPMENT	*39*
SOCIAL BEHAVIOUR IN ANIMALS REPRODUCTION	*40*
SEX-LINKED INHERITANCE	*40*
HAEMOPHILIA A	*42*
DUCHENNE'S MUSCULAR DYSTROPHY (DMD)	*43*
UNISEXUALITY	*43*
SEX MATING	*44*
IMPOTENCE	*45*
SEX THERAPY	*45*
BISEXUALITY	*46*
HERMAPHRODITISM	*46*

PSYCHOMOTOR LEARNING IN SEX	47
CROSS-FERTILIZATION	48
HUMAN DEVELOPMENT	48
REPRODUCTIVE BEHAVIOUR – DISPLAYS	51
ARTHROPOD - REPRODUCTIVE SYSTEM AND LIFE CYCLE	52
3. HUMAN BEHAVIOUR	55
SOCIAL AND CULTURAL ASPECTS	55
PSYCHOLOGICAL EFFECTS OF EARLY CONDITIONING	57
NERVOUS SYSTEM FACTORS	60
SEXUALLY TRANSMITTED DISEASES	61
IMPACT OF PSYCHOANALYSIS	62
PSYCHOSEXUAL DYSFUNCTION	63
GENETIC AND CONGENITAL ABNORMALITIES	64
ANARCHISM	65
PERSONALITY	66
CULTURE AND PERSONALITY	67
MARITAL CUSTOMS AND LAWS	68
MARITAL ROLES	70
PEER SOCIALIZATION	71
PROBLEMS IN DEVELOPMENT	71
4. SEX AND DEVELOPMENTAL PSYCHOLOGY	73
BABY AND CHILDHOOD SEXUALITY	73
EARLY ADOLESCENCE	73
MID-AND LATE ADOLESCENCE	74
SEXUAL ATTRACTION	74
SEXUAL ENCOUNTERS	74
ROMANCE	75
PARTNERSHIP/MARRIAGE	75
SEXUAL ANATOMY	76
COPULATION	76
TEACHING CHILDREN ABOUT SEX	77
MASTURBATION	77
HOMOSEXUALITY	77
SEX IN OLD AGE	77
LOVE	78
5. IDEALISTIC HUMAN REARING	79
RAISING A CHILD	79
CHILD CULTURE	80
PHYSICAL ASPECTS	85
PSYCHOLOGICAL FACTORS	87
PREGNANCY PERIOD	92
CHILDREN'S IDEALS	93
HEREDITY AND PATERNAL INFLUENCE	96
DAY OF BIRTH	99
CARE FOR THE NEWBORN CHILD	104
POWER OF WORDS	107
ANSWERING QUESTIONS	109
OVERCOMING DIFFICULTIES	114
CHILD'S HEALTH	116
HABIT FORMATION AND PUNISHMENT	117
PERIOD OF TRANSITION	120
RESPONSIBILITIES REVIEWED	123
INDEX	125
BIBLIOGRAPHY	127

'Eros punished by Venus,' fresco from Pompeii

SEX, AN EXPLORATION OF SEXUALITY, EROS AND LOVE

PREFACE

Sexuality is a vast subject covering many fields of study. Genitality, with which it is usually confused, is only small part of it, yet sex books and sex education tend to deal almost entirely with sexuality.

This book describes sexuality as a new perspective on love, emotion and relationships. The contents play such an important part in sex for every member of the family.

By studying and discussing human and sometimes animal sexuality from infancy onwards, this book will offer a better understanding of sex and its many diverse situations, starting with the mythological comprehension of Eros (the god of love), including sex as a biological and physiological subject, extending into the psychology of sex and concluding with the realm of idealistic child rearing.

In religion, morality, social customs and artistic themes, sex is and has always been an immensely powerful factor. In fact, sex pervades every culture in the world.

1. MYTHOLOGICAL EROS
EROS, CUPID AND SEX

The Anglo-Saxon word of sex is very closely associated with the mythological Eros, the Greek god of love, a primeval force, who as various handed down stories explain, he was the son of Aphrodite. In Hellenistic times he became associated with romantic love, and was represented as a little winged archer, shooting his arrows at gods and men. The Romans identified him with Cupid.

In Greek religion, he was the god of love. In the *Theogony* of Hesiod (fl. 700 BC), Eros was a primeval god, son of Chaos, the original primeval emptiness of the universe; but later tradition made him the son of Aphrodite, goddess of sexual love and beauty, by either Zeus (the king of the gods), Ares (god of war and of battle), or Hermes (divine messenger of the gods). Eros was god not simply of passion but also of fertility. His brother was Anteros, the god of mutual love, who was sometimes described as his opponent.

The chief associates of Eros were Pathos and Himeros (Longing and Desire). Later writers assumed the existence of a number of Eroses (like the several versions of the Roman Amor). In Alexandrian poetry he degenerated into a mischievous child. In archaic art he was represented as a beautiful winged youth but tended to be made younger and younger until, by the Hellenistic period, he was an infant. His chief cult centre was at Thespiae in Boeotia, where the Erotidia were celebrated. He also shared a sanctuary with Aphrodite on the north wall of the Acropolis at Athens.

All in all, Eros specialised in making people fall in love, even non-suitable couples, he was considered to be mischievous, fond of romantic intrigues.

On the other hand, Cupid the ancient Roman god of love in all its varieties, was the counterpart of the Greek god Eros and the equivalent of Amor in Latin poetry. According to myth, Cupid was the son of Mercury, the winged messenger of the gods, and Venus, the goddess of love; he usually appeared as a winged infant carrying a bow and a quiver of arrows, whose wounds inspired love or passion in his every victim. He was sometimes portrayed wearing armour like that of Mars, the god of war, perhaps to suggest ironic parallels between warfare and romance or to symbolize the invincibility of love.

Although some literature portrayed Cupid as callous and careless, he was generally viewed as beneficent, on account of the happiness he imparted to couples both mortal and immortal. At the worst he was considered mischievous in his matchmaking, this mischief often directed by his mother, Venus. In one tale, her machinations backfired when she used Cupid in revenge on the mortal Psyche, only to have Cupid fall in love and succeed in making Psyche his immortal wife.

Various famous people brought forward various other versions. Freud, for instance, suggested that even Eros is not fully in harmony with civilization, for the libidinal ties creating collective solidarity are aim-inhibited and diffuse rather than directly sexual. Thus, there is likely to be tension between the urge for sexual gratification and the sublimated love for mankind.

Furthermore, because Eros and Thanatos are themselves at odds, conflict and the guilt it engenders are virtually inevitable. The best to be hoped for is a life in which the repressive burdens of civilization are in rough balance with the realization of instinctual gratification and the sublimated love for mankind. But reconciliation of nature and culture is impossible, for the price of any civilization is the guilt produced by the necessary thwarting of man's instinctual drives.

Although elsewhere Freud had postulated mature, heterosexual genitality and the capacity to work productively as the hallmarks of health and urged that "where id is, there shall ego be," it is clear that he held out no hope for any collective relief from the discontents of civilization. He only offered an ethic of resigned authenticity, which taught the wisdom of living without the possibility of redemption, either religious or secular.

The Libido concept originated by Sigmund Freud signifies the instinctual physiological or psychic energy associated with sexual urges and, in his later writings, with all constructive human activity. In the latter sense of eros, or life instinct, libido was opposed by Thanatos, the death instinct and source of destructive urges; the interaction of the two produced all the variations of human activity. Freud considered psychiatric symptoms the result of misdirection or inadequate discharge of libido.

Carl Jung used the term in a more expansive sense, encompassing all life processes in all species. Later theories of motivation have substituted for libido such related terms as drive and tension.

SYMBOLISM OF SEX AND THE LIFE CYCLE

The symbols of sexuality and the life cycle perform a function similar to those of time and eternity in the higher religions. They indicate the permanence of the cycle of sexual functions and the return and renewal of individual and collective physical life. The endless renewal of life is variously represented. It may be as realistic depictions or diagrammatic and stylized abbreviations of man and woman, god and goddess, masculine and feminine animals in the act of love and sexual union, as in reliefs on Hindu temples, or as depictions of sex characteristics (*e.g.*, in Indian *linga-yoni* symbolism).

The theme of renewal also may be depicted in representations of woman with emphasis on her function as mother, as in the nursing-mother figures of ancient Greece. The life cycle also is represented by figures portraying the ages of man or by depictions of pain and suffering, as in pictures of the Buddha's death, which also indicate his breaking out of the endless chain of existence.

SEX, SEXUALITY AND REPRODUCTION

In many biological writings, sex is the sum of features by which members of species can be divided into two groups--male and female-- that complement each other reproductively.

Sex, sexuality, and reproduction are all closely woven into the fabric of living things. All relate to the propagation of the race and the survival of the species. Yet there can be sex without sexuality, and reproduction need not be sexual, although for most forms of life sexual reproduction is essential for both propagation and long-term survival.

LEGAL REGULATION

Sex laws, the origins of which are found within the church, are unique in one important respect. Whereas all other laws are basically concerned with the protection of person or property, the majority of sex laws are concerned solely with maintaining morality. The issue of morality is minimal in other laws: one can legitimately evict an impoverished old couple from their mortgaged home or sentence a hungry man for stealing food. Only in the realm of sex is there a consistent body of law upholding morality.

The earliest sex laws of which there is knowledge are from the Near East and date back to the 2nd millennium BC. They are remarkable in three respects: there are great omissions--certain acts are not mentioned whereas others receive detailed attention; some laws seem almost contradictory; and penalties are often extraordinarily severe. One obtains the distinct impression that these laws were case law--that is, laws formulated upon specific cases as they arose rather than being the result of lengthy judicial deliberation done in advance. These laws influenced Judaic and, hence, Christian thinking, and some were immortalized in the Bible, chiefly in Leviticus.

When secular law replaced religious law, there was rather little change in content. In Europe the Napoleonic Code represented a break with tradition and introduced some measure of sexual tolerance, but in England and the United States there was no such rift with the past. In the latter country, as each new state joined the union, its sex laws simply duplicated, to a great extent, those of pre-existing states; legislators were disinclined to debate sexual issues or to risk losing votes by discarding or weakening sex laws.

Sex laws may be grouped in three categories:

(1) Those concerned with protection of person. These are based on the element of consent. These otherwise logical laws become problematic when society deems that minors, mental retardates, and the insane are incapable of giving consent--hence, coitus with them is rape.

(2) Those concerned with preventing offence to public sensibilities. Statutes preclude public sexual activity, exhibitionism, and offensive solicitation.

(3) Those concerned with maintaining sexual morality. These constitute the majority of sex laws, covering such items as premarital coitus, extramarital coitus, incest, homosexuality, prostitution, peeping, nudity, animal contact, transvestism, censorship, and even specific sexual techniques--chiefly oral or anal.

Laws relating to sexual conduct and morality are generally far more extensive in the United States than in Western Europe and most other areas of the world.

In recent years, in Europe and the United States, a number of highly respected legal, medical, and religious organizations have deliberated on the issue of the legal control of human sexuality. They have been unanimous in the conclusion that, while laws protecting person and public sensibilities should be retained, the purely moral laws should be dropped. What consenting adults do in private, it is argued, should not be subject to legal control.

In the final analysis, sexuality, like any other vital aspect of human life, must be dealt with on an individual or societal level with a combination of rationality, sensitivity, and tolerance if society is to avoid personal and social problems arising from ignorance and misconception.

SEXUAL BEHAVIOUR

Human sexual behaviour refers to any activity--solitary, between two persons, or in a group--that induces sexual arousal. There are two major determinants of human sexual behaviour: the inherited sexual response patterns that have evolved as a means of ensuring reproduction and that are a part of each individual's genetic inheritance, and the degree of restraint or other types of influence exerted on the individual by society in the expression of his sexuality. The objective here is to describe and explain both sets of factors and their interaction.

It should be noted that taboos in Western culture and the immaturity of the social sciences for a long time impeded research concerning human sexual behaviour, so that by the early 20th century scientific knowledge was largely restricted to individual case histories that had been studied by such European writers as Sigmund Freud, Havelock Ellis, and Richard, Freiherr von Krafft-Ebing.

By the 1920s, however, the foundations had been laid for the more extensive statistical studies that were conducted before World War II in the United States. Of the two major organizations for sex study, one, the Institut für Sexualwissenschaft in Berlin (established in 1897), was destroyed by the Nazis in 1933. The other, the Institute for Sex Research, begun in 1938 by the American sexologist Alfred Charles Kinsey at Indiana University, Bloomington, undertook the study of many aspects of human sexual behaviour. Much of the following discussion rests on the findings of the Institute for Sex Research, which comprise the most comprehensive data available. The only other country for which comprehensive data exist is Sweden.

FOLK ART AND VARIATIONS BY SEX

In modern square dancing the difference between male and female styles is negligible, but in most folk dances the women move more gently than the men, with smaller steps, lower leaps, and less raising of the knees or feet. The women dancers have a more sinuous, alluring style in southern Spain; they spin gently in the Austrian and Bavarian *Schuhplattler* and the Caucasian *lezginka*, while the men jump, clap, and shout. Among American Indian tribes the women have a more subdued style and often special, tiny steps except in couple dances that have been adapted from the mainstream of Western social dancing.

ECOLOGICAL INFLUENCES

The setting affects the movement style. Joan Lawson suggests differences due to the natural environment--a theory that will need more investigation. She maintains that in rich agricultural plains or river valleys, such as the Danubian Plains and parts of France, and Denmark, movements are accented downward as if the body were being drawn toward the soil. Dancers perform in large groups, using the same step, closely linked together by fingers, hands, elbows, or shoulders. By contrast, in mountainous areas there is a good deal of leaping and individual display, especially among the males.

2. BIOLOGICAL AND PHYSIOLOGICAL ASPECTS OF SEX
SEXUAL AND NON-SEXUAL REPRODUCTION

Because the life span of all individual forms of life, from microbes to man, is limited, the first concern of any particular population is to produce successors. This is reproduction, pure and simple. Among lower animals and plants it may be accomplished without involving eggs and sperm. Ferns, for example, shed millions of microscopic, nonsexual spores, which are capable of growing into new plants if they settle in a suitable environment. Many higher plants also reproduce by nonsexual means. Bulbs bud off new bulbs from the side.

Certain jellyfish, sea anemones, marine worms, and other lowly creatures bud off parts of the body during one season or another, each thereby giving rise to populations of new, though identical, individuals. At the microscopic level, single-celled organisms reproduce continually by growing and dividing successively to give rise to enormous populations of mostly identical descendants. All such reproduction depends on the capacity of cells to grow and divide, which is a basic property of life. In the case of most animals, however, particularly the higher forms, reproduction by nonsexual means is apparently incompatible with the structural complexity and activity of the individual.

Although nonsexual reproduction is exploited by some creatures to produce very large populations under certain circumstances, it is of limited value in terms of providing the variability necessary for adaptive advantages. Such so-called vegetative forms of reproduction, whether of animals or plants, result in individuals that are genetically identical with the parent. If some adverse environmental change should occur, all would be equally affected and none might survive. At the best, therefore, nonsexual reproduction can be a valuable and perhaps an essential means of propagation, but it does not exclude the need for sexual reproduction.

Sexual reproduction not only takes care of the need for replacement of individuals within a population but gives rise to populations better suited to survive under changing circumstances. In effect it is a kind of double assurance that the race or species will persist for an indefinite time. The great difference between the two types of reproduction is that individual organisms resulting from nonsexual reproduction have but a single parent and are essentially alike, whereas those resulting from sexual reproduction have two parents and are never exact replicas of either.

Sexual reproduction thus introduces a variability, in addition to its propagative function. Both types of reproduction represent the capacity of individual cells to develop into whole organisms, given suitable circumstances. Sex is therefore something that has been combined with this primary function and is responsible for the capacity of a race to adapt to new environmental conditions.

ASEXUAL REPRODUCTION

Both homosporous and heterosporous life histories may exhibit various types of asexual reproduction (vegetative reproduction, somatic reproduction). Asexual reproduction is any reproductive process that does not involve meiosis or the union of nuclei, sex cells, or sex organs. Depending on the type of life history, asexual reproduction can involve the $1n$ or $2n$ generation.

The significance of sexual reproduction is that it is responsible for the genetic variation arising in a population as a result of the segregation and recombination of genetic material via meiosis and syngamy, respectively (the cells that result from sexual reproduction are genetically different from their parent cells). The significance of asexual reproduction is that it is a means for a rapid and significant increase in the numbers of individuals. (Weeds, for instance, are successful partly due to their great capacity for vegetative reproduction.)

The cells that result from asexual reproduction are genetically identical to their parent cells. In addition, vegetative reproduction in the bryophytes and pteridophytes is a means of bypassing the somewhat lengthy and moisture-dependent sexual process--that is, the motile, swimming sperm characteristic of these groups require the presence of water, which may be a limiting factor in drier times.

SEX DETERMINATION

Sex determination is the term used for the sex of an organism, as decided by environmental or genetic factors. For example, if embryogenesis in the map turtle occurs below 28o C only males hatch but if above 30o C only females.

In most higher animals, including man, sex is genetically controlled. In mammals the presence or absence of the Y chromosome determines sex, thus XX individuals are female and XY male. A single gene on the Y chromosome triggers off a cascade of effects which leads to the differentiation of male gonads and genitalia. The absence of this gene leads to female development.

SEX DIFFERENCES

Sex differences (the term use in psychology), relates to the differences in males and females. Statistics showing sex differences are often unreliable, since there may be differential reporting of abilities and disabilities between the sexes. However, it has been repeatedly shown that females have greater verbal skills and males greater visual-spatial and mathematical skills.

The origin of these differences in genetic or environmental factors continues to be hotly disputed. Temperamental differences, especially in the display of aggression, are often linked to hormones and to the primary and secondary sexual characteristics important for reproduction.

There are well-documented sex differences in the age of reaching puberty, in total body growth, in longevity, and in susceptibility to neuroses and depression. No fully satisfactory theory exists to explain how such biological differences interact with social processes to give rise to tertiary level characteristics, such as the lesser representation of females among mathematicians or in the sciences.

Differential growth rates may result in a variance in the patterns of brain differentiation for verbal and spatial skills. The distribution of such abilities between the sexes, although overlapping, would then show a different range. Sex roles within the social system (see gender) may further exaggerate such inherent sex differences as children learn through socialization to take their place in the social world.

SEX CELLS

The term sex is variously employed. In the broad sense it includes everything from the sex cells to sexual behaviour. Primary sex, which is generally all that distinguishes one kind of individual from another in the case of many lower animals, denotes the capacity of the reproductive gland, or gonad, to produce either sperm cells or eggs or both. If only sperm cells are produced, the reproductive gland is a testis, and the primary sex of the tissue and the individual possessing it is male. If only eggs are produced, the reproductive gland is an ovary, and the primary sex is female. If the gland produces both sperm and eggs, either simultaneously or successively, the condition is known as hermaphroditic. An individual, therefore, is male or female or hermaphrodite primarily according to the nature of the gonad.

As a rule, male and female complement each other at all levels of organization: as sex cells; as individuals with either testes or ovaries; and as individuals with anatomical, physiological, and behavioural differences associated with the complemental roles they play during the whole reproductive process. The role of the male individual is to deliver sperm cells in enormous numbers in the right place and at the right time to fertilize eggs of female individuals of the same species. The role of the female individual is to deliver or otherwise offer eggs capable of being fertilized under precise circumstances.

In the case of hermaphrodite organisms, animal or plant, various devices are employed to ensure cross-fertilization, or cross-pollination, so that full advantage of double parentage is obtained. The basic requirement of sexual reproduction is that reproductive cells of different parentage come together and fuse in pairs. Such cells will be genetically different to a significant degree, and it is this feature that is essential to the long-term well-being of the race. The other sexual distinctions, between the two types of sex cell and between two individuals of different sex, are secondary differences connected with ways and means of attaining the end.

SEX CHROMOSOMES

In most species of animals the sex of individuals is determined decisively at the time of fertilization of the egg, by means of chromosomal distribution. This process is the most clear-cut form of sex determination. When any cell in the body divides, except during the formation of the sex cells, each daughter cell receives the full complement of chromosomes; *i.e.*, copies of the two sets of chromosomes derived from the sperm cell and egg, respectively.

The two sets are similar except for one pair of chromosomes. These are the so-called sex chromosomes, and the pair may be exactly alike or they may be obviously different, depending on the sex of the individual. The sex chromosomes are of two types, which are designated X and Y, and the pair of sex chromosomes may consist of two X chromosomes or of an X and Y paired together.

In mammals (including man) and flies, the cells of males contain an XY pair and the cells of females contain an XX pair. On the other hand, in butterflies, fishes, and birds, the cells of females contain an XY pair and those of males contain an XX pair. In either case the Y chromosome is generally smaller than the X chromosome and may even be absent.

What is most important concerning chromosomal sex determination is whether the cells of the individual contain one X chromosome or two X chromosomes. Human beings, for example, have cells with 22 pairs of nonsexual chromosomes, or autosomes, together with an XX pair or an XY pair. The female has a total of 46 functional chromosomes; the male has 45 plus a Y, which is mainly inert. Sex determination thus becomes a matter of balance. With one X chromosome plus the 44 autosomes in every cell, the whole course of development of primary and secondary sexual characteristics is toward the male; with two X chromosomes plus the autosomes in every cell, the whole system is swung over to the female.

The manipulation of this control system is readily accomplished during the special process of cell division that takes place in the gonads to produce sperm and eggs and their subsequent union at fertilization. In mammals, for example, since all cells in the female contain two X chromosomes, all the eggs will receive a single X chromosome when they are formed. All eggs are accordingly the same in this respect. In contrast, all cells in the male have the XY constitution, and therefore, when the double set of chromosomes is reduced to a single set during the formation of the spermatozoa, half of the spermatozoa will receive an X and half will receive a Y. Consequently, when an egg is fertilized by a sperm, the chances are about equal that the sperm will carry an X or will carry a Y, since the two types are inevitably produced in equal numbers. If it carries an X, the XX female constitution results; if a Y, then the XY male constitution results.

ABNORMAL CHROMOSOME EFFECTS

Occasionally, however, the processes of chromosomal re-assortment and recombination occurring during sex cell formation and fertilization depart somewhat from the normal course. Sperm and eggs may be produced that are oversupplied or undersupplied with sex chromosomes. Fertilized eggs in humans may, for instance, have abnormal sex chromosome constitutions such as XXX, XXY, or XO. Those with the triple-X chromosome constitution have all the appearance of normal females and are called, in fact, super-females, although only some will be fertile. Those with the XO (one X, but lacking Y altogether) constitution, a much more common condition, are also feminine in body form and type of reproduction system but remain immature. Individuals with the XXY constitution are outwardly males but have small testes and produce no spermatozoa. Those with the more abnormal and relatively rarer constitutions XXXXY and XXYY are typically mentally defective and in the latter case are hard to manage. Thus abnormal combinations generally result in an infertility on the one hand and an abnormal sexuality in the whole system, for either too little or too much of what is ordinarily good can be disastrous.

Very different kinds of abnormal development resulting from faulty chromosomal distribution are particularly observable in insects. The most common form in flies is an individual that is male on one side, female on the other, with a sharp line of demarcation. In other cases one-quarter of the body may be male and three-quarters female, or the head may be female and the rest of the body, male. These types are known as gynandromorphs, or sexual mosaics, and result from aberration in the distribution of the X chromosomes among the first cells to be formed during the early development of the embryo. This condition is unknown among higher animals.

PSEUDOHERMAPHRODITISM

in human beings, a condition in which the individual has a single chromosomal and gonadal sex but combines features of both sexes in the external genitalia, causing doubt as to the true sex. Female pseudohermaphroditism refers to an individual with ovaries but with secondary sexual characteristics or external genitalia resembling those of a male. Usually at puberty the female secondary sex characteristics develop. If the condition is identified at birth, the child may be raised as a female with a minimum of social readjustment. Administration of certain corticosteroids prevents further development of the condition, and surgery may be used to correct any residual genital defects.

Male pseudohermaphroditism refers to individuals whose gonads are testes but whose secondary sexual characteristics or external genitalia resemble those of a female. In this disorder the foetal target organs are unable, for unknown reasons, to react to testosterone produced by the foetal testes. The most common type is testicular feminization, wherein the external genitals are entirely feminine, and at puberty female secondary sex characteristics appear, yet the gonads are testes, and the sex-chromosome pattern is male. The disorder is sometimes recognized at puberty when menstruation fails to begin. In testicular feminization, because there is little or no response to male hormones and because the external genitalia are female, the child is raised as a female. Other forms of male pseudohermaphroditism can be altered to become "complete" male, and such children may be raised as males.

TRANSSEXUALISM

Transsexualism is the term used to explain the 'disturbance' of gender identity in which the affected person believes that he or she should belong to the opposite sex. The transsexual male, for example, is born with normal male genitalia and other secondary characteristics of the masculine sex; very early in life, however, he identifies with women and behaves in a manner appropriate to the female sex. His sexual orientation is generally one of attraction to other males. Unlike the male homosexual, however, the transsexual male desires to relate to other men and be treated by them as if he were a normal female.

With the development of successful surgical techniques and hormone therapy, several thousand transsexuals, male and female, have undergone a permanent sex change. Although both male and female transsexuals exist, the male-to-female operation is more common because the genital reconstruction is more satisfactory. The male transsexual's penis and testes are removed and an artificial vagina is created; breast implants may be inserted, although some breast development usually is promoted with the use of feminizing hormones. Female transsexuals may undergo mastectomy and hormone treatments to produce the male secondary sexual characteristics, but attempts to create an artificial penis have not been particularly satisfactory.

SEX DIFFERENCES IN ANIMALS

In many animals, sexual differences are apparent in addition to the primary sex differentiation into males with testes and females with ovaries and apart from the accessory structures and tissues associated with the presence of one kind of sex gland or the other. Secondary sex differentiation in sexually distinct individuals is to be seen in many forms. In humans, for example, the beard and deep voice of the male and the enlarged breasts of the female are features of this sort.

The great claw of the fiddler crab, the antlers of a moose, the great bulk and strength of a harem master in a fur seal colony, the beautiful fan tail of the peacock, and the bright feathers of other birds, are all distinctively male characteristics, and all are associated with the sexual drive of males. Females, by and large, are of comparatively quiet disposition and relatively drab appearance. Their function is to produce and nurture eggs, as safely and usually as inconspicuously as possible. The male function is to find and fertilize the female, for which both drive and display are generally required.

It is the business of sperm to be active and so find an egg. Similarly it is the business of males to find a female and mate with her if possible. The male drive, or male eagerness, is a consequence of this special function of males. In nature, males possessing a strong eagerness to mate will find more females and leave more progeny than males lacking in sex drive. The progeny moreover will tend to inherit the drive of the parent.

Males therefore are generally competitive with other males, with a premium placed on physical strength and sex drive and also on various devices for the attraction and stimulation of the female. The various exclusively male features already listed are all examples of characteristics of this sort, and they are related to the securing of female mates rather than the actual fertilization of eggs or to the problems of survival and adaptation.

SYSTEM, IN FROG, MOUSE, AND MAN ALIKE

In the young embryo a pair of gonads develop that are indifferent or neutral, showing no indication whether they are destined to develop into testes or ovaries. There are also two different duct systems, one of which can develop into the female system of oviducts and related apparatus and the other into the male sperm duct system. As development of the embryo proceeds, either the male or the female reproductive tissue differentiates in the originally neutral gonad of the mammal.

In the frog and other lower vertebrate animals, the picture is even clearer. The original gonad consists of an outer layer of cells and an inner core of cells. If the individual is to be a male, the central tissue grows at the expense of the outer layer. If it is to be a female, the outer tissue grows at the expense of the central core tissue. If both should grow, which is a possibility although a rare occurrence, the individual will be a hermaphrodite. Anything that influences the direction taken therefore may be said to determine sex.

SEX DETERMINATION

The determination of the sex of an individual, with regard to both the primary sex--*i.e.*, whether the ovaries or the testes develop--and the various secondary sexual characteristics may be rigorously controlled from the start of development or may be subject to later influences of a hormonal or environmental nature. However this may be, in order to appreciate the action of the control systems, the point of departure is that animals were primitively hermaphrodite, that during early stages of evolution every individual probably possessed both male and female gonads.

Differentiation into separate sexes, each possessing male or female gonads but not both at the same time, is a device to ensure cross-fertilization of eggs, whether this is accomplished by having the two types of sexual gland mature at different stages of the growth of the individual, as in some shrimp and others, or whether by the production of two distinct types of individuals, as in most species of animals.

This point of view is important because the question ceases to be how testes are caused to develop in the male organism and ovaries in the female but how, in a potentially double-sexed organism, the development of one or the other sex is suppressed. That such is the case is seen as clearly as anywhere in the human condition itself. Neither sex is completely male or female. Females have functional, well-developed mammary glands. Males also have mammary glands, undeveloped and non-functional although equipped with nipples. Males have a penis for delivering sperm, but females have a small, non-functional equivalent-- the clitoris. These are secondary sexual features, to be sure, but the difference between the sexes is in the degree of their development, not a matter of absolute presence or absence.

The basis for this is seen in the very beginnings of the development of the reproductive system, in frog, mouse, and man alike. In the young embryo a pair of gonads develop that are indifferent or neutral, showing no indication whether they are destined to develop into testes or ovaries. There are also two different duct systems, one of which can develop into the female system of oviducts and related apparatus and the other into the male sperm duct system. As development of the embryo proceeds, either the male or the female reproductive tissue differentiates in the originally neutral gonad of the mammal.

In the frog and other lower vertebrate animals, the picture is even clearer. The original gonad consists of an outer layer of cells and an inner core of cells. If the individual is to be a male, the central tissue grows at the expense of the outer layer. If it is to be a female, the outer tissue grows at the expense of the central core tissue. If both should grow, which is a possibility although a rare occurrence, the individual will be a hermaphrodite. Anything that influences the direction taken therefore may be said to determine sex.

SEX PATTERNS

Since the great value of sex as distinct from reproduction is the re-assortment and recombination of genes every generation, sex cells from two separate parents ordinarily give rise to the greatest variation, unless the parental individuals are themselves too closely related to each other. The presence of male and female individuals, respectively, generally produced in approximately equal numbers, is characteristic of so much of the animal kingdom that it appears to be the natural state.

All that is certain, however, is that this condition has evolved as the most effective means to the particular end, and it may have done so independently among the various more or less unrelated groups of animals. The condition of separate sexes is not a universal fact, and two sexes within the same individual is typical of the more sluggish or actually attached kinds of animal life. Earthworms, slugs, land snails, flatworms, tapeworms, barnacles, sea squirts, and some others are all double-sexed individuals, or hermaphrodites. All have ovaries and testes producing mature eggs and sperm at the same time.

Nevertheless, cross-fertilization is accomplished, and self-fertilization, even though possible, is generally avoided. Of those kinds of animal life mentioned above, all except the sea squirts have well-encased eggs that need to be fertilized before being laid. Mutual copulation, whereby each member of a mating pair of individuals introduces sperm into the body of the other member, is characteristic of these creatures, with the exception of the sea squirts.

When animals shed sperm and comparatively naked eggs into the surrounding water, as is the case in sea squirts, self-fertilization is difficult to avoid. Most creatures have evolved an effective separation of the sexes between different individuals. Even so, there are more ways than one of accomplishing this. The common means is to produce male and female individuals that are constitutionally different, yet an equally effective procedure is for all individuals to be constitutionally the same but to become mature as male or female at different stages of the growth cycle.

The oyster on its rock changes sex from male to female and back again once or twice a year. Certain shrimps also are hermaphrodites. Each young shrimp of this kind grows up to be a male and is fully and functionally a male when about half the size of the females. As the next season approaches, his testes shrink, no more spermatozoa are produced, and ovaries begin to enlarge. As full growth is reached, the shrimp that had been a male becomes a typical female, ready to mate again, but this time with a young male of a newer generation. The system works as well as any other and clearly has its points. In fact the hagfish, not a true fish but a more primitive jawless vertebrate, also changes sex regularly, from year to year.

SEX HORMONES

Sex hormones are a chemical substance produced by a sex gland or other organ that has an effect on the sexual features of an organism. Like many other kinds of hormones, sex hormones may also be artificially synthesized.

PRODUCTION

Androgen - any of a group of hormones that primarily influence the growth and development of the male reproductive system.

The predominant and most active androgen is testosterone, which is produced by the male testes. The other androgens, which support the functions of testosterone, are produced mainly by the adrenal cortex--the outer substance of the adrenal glands--and only in relatively small quantities. Trace quantities of androgens are found in the female blood plasma. It is believed that the adrenal glands produce most of these small amounts. The ovaries, which normally secrete the female hormones known as estrogens, also produce minute amounts of androgens.

In the male, the interstitial cells of Leydig, located in the connective tissue surrounding the sperm-producing tubules of the testes, are responsible for the production and secretion of androgens. In male animals that breed only seasonally, such as migratory birds and sheep, Leydig cells are prevalent in the testes during the breeding season but diminish considerably in number during the non-breeding season. The actual secretion of androgens by these cells is controlled by luteinizing hormone (LH) from the pituitary gland.

OESTROGEN

Oestrogen is any of a group of hormones that primarily influence the female reproductive tract in its development, maturation, and function. There are three major hormones--estradiol, estrone, and estriol--among the estrogens, estradiol being the predominant one. The major sources of estrogens are the ovaries and the placenta (the temporary organ that serves to nourish the foetus and remove its wastes); additional small amounts are secreted by the adrenal glands and by the male testes.

It is believed that the egg follicle (the saclike structure that holds the immature egg) and interstitial cells (certain cells in the framework of connective tissue) in the ovary are the actual production sites of estrogens in the female. Oestrogen levels in the bloodstream seem to be highest during the egg-releasing period (ovulation) and after menstruation, when tissue called the corpus luteum replaces the empty egg follicle.

Oestrogens affect the ovaries, vagina, fallopian tubes, uterus, and mammary glands. In the ovaries, estrogens help to stimulate the growth of the egg follicle; they also stimulate the pituitary gland in the brain to release hormones that assist in follicular development. Once the egg is released, it travels through the fallopian tubes on its way to the uterus;

in the fallopian tubes estrogens are responsible for developing a thick muscular wall and for the contractions that transport the egg and sperm cells.

The young female uterus, if deprived of estrogens, does not develop into its adult form; the adult uterus that does not receive estrogens begins to show tissue degeneration. Estrogens essentially build and maintain the endometrium--a mucous membrane that lines the uterus; they increase the endometrium's size and weight, cell number, cell types, blood flow, protein content, and enzyme activity.

Oestrogens also stimulate the muscles in the uterus to develop and contract; contractions are important in helping the wall to slough off dead tissue during menstruation and during the delivery of a child and of the placenta. The cervix, the tip of the uterus, which projects into the vagina, secretes mucus that enhances sperm transport; estrogens are thought to regulate the flow and thickness of the mucous secretions. The growth of the vagina to its adult size, the thickening of the vaginal wall, and the increase in vaginal acidity that reduces bacterial infections are also correlated to oestrogen activities.

In the breasts the actions of estrogens are complexly interrelated with those of other hormones, and their total significance is not easily defined; they are, however, responsible for growth of the breasts during adolescence, pigmentation of the nipples, and the eventual cessation of the flow of milk.

Oestrogens also influence the structural differences between the male and female bodies. Usually the female bones are smaller and shorter, the pelvis is broader, and the shoulders are narrower. The female body is more curved and contoured because of fatty tissue that covers the muscles, breasts, buttocks, hips, and thighs. The body hair is finer and less pronounced, and the scalp hair is usually more permanent. The voice box is smaller and the vocal cords shorter, giving a higher-pitched voice in females than in males. In addition, oestrogens suppress the activity of sebaceous (oil-producing) glands and thereby reduce the likelihood of acne in the female. In experimental animals, loss of estrogens diminishes the mating desires and other sexual behaviour patterns.

In the male, traces of estrogens are present in the blood and urine; estrogens seem to be most evident in the male during puberty and old age. Their function in the male and their interplay with the male hormones are not completely known.

FERTILIZATION, SEX DETERMINATION, AND DIFFERENTIATION

A human individual arises through the union of two cells, an egg from the mother and a sperm from the father. Human egg cells are barely visible to the naked eye. They are shed, usually one at a time, from the ovary into the oviducts (fallopian tubes), through which they pass into the uterus. Fertilization, the penetrance of an egg by a sperm, occurs in the oviducts. This is the main event of sexual reproduction and determines the genetic constitution of the new individual.

Human sex determination is a genetic process that depends basically on the presence of the Y chromosome in the fertilized egg. This chromosome stimulates a change in the undifferentiated gonad into that of the male (a testicle). The gonadal action of the Y chromosome is mediated by a gene located near the centromere; this gene codes for the production of a cell surface molecule called the H-Y antigen. Further development of the anatomic structures, both internal and external, that are associated with maleness is controlled by hormones produced by the testicle. The sex of an individual can be thought of in three different contexts: chromosomal sex, gonadal sex, and anatomic sex. Discrepancies among these, especially the latter two, result in the development of individuals with ambiguous sex, often called hermaphrodites. The phenomenon of homosexuality is of uncertain cause and is unrelated to the above sex-determining factors. It is of interest that in the absence of a male gonad (testicle) the internal and external sex anatomy is always female, even in the absence of a female ovary. A female without ovaries will, of course, be infertile and will not experience any of the female developmental changes normally associated with puberty. Such a female will often have Turner's syndrome.

If X-containing and Y-containing sperm are produced in equal numbers, then according to simple chance one would expect the sex ratio at conception (fertilization) to be half boys and half girls, or 1 : 1. Direct observation of sex ratios among newly fertilized human eggs is not yet feasible, and sex-ratio data are usually collected at the time of birth. In almost all human populations of newborns there is a slight excess of males; about 106 boys are born for each 100 girls. Throughout life, however, there is a slightly greater mortality of males; this slowly alters the sex ratio until, beyond the age of about 50 years, there is an excess of females. Studies indicate that male embryos suffer a relatively greater degree of prenatal mortality, so that the sex ratio at conception might be expected to favour males even more than the 106 : 100 ratio observed at birth would suggest. Firm explanations for the apparent excess of male conceptions have not been established; it is possible that Y-containing sperm survive better within the female reproductive tract, or that they may be a little more successful in reaching the egg in order to fertilize it. In any case, the sex differences are small, the statistical expectation for a boy (or girl) at any single birth still being close to one out of two.

During gestation--the period of nine months between fertilization and the birth of the infant--a remarkable series of developmental changes occur. Through the process of mitosis, the total number of cells changes from one (the fertilized egg) to about 2×10^{11}. In addition, these cells differentiate into hundreds of different types with specific functions (liver cells, nerve cells, muscle cells, etc.). A multitude of regulatory processes, both genetically and environmentally controlled, accomplish this differentiation. Elucidation of the exquisite timing of these processes remains one of the great challenges of human biology.

HISTORY OF SEX HORMONES

Not the least of the advances in endocrinology was the increasing knowledge and understanding of the sex hormones. This culminated in the application of this knowledge to the problem of birth control. After an initial stage of hesitancy, the contraceptive pill, with its basic rationale of preventing ovulation, was accepted by the vast majority of family-planning organizations and many gynaecologists as the most satisfactory method of contraception. Its risks, practical and theoretical, introduced a note of caution, but this was not sufficient to detract from the wide appeal induced by its effectiveness and ease of use.

CHEMICAL COMPOUND OF SEX HORMONES

Steroids that have a phenolic ring A (*i.e.,* those in which ring A is aromatic and bears a hydroxyl group) are ubiquitous products of the ovary of vertebrate animals. These are the estrogens, of which estradiol is the most potent. They maintain the female reproductive tissues in a fully functional condition, promote the oestrous state of preparedness for mating, and stimulate development of the mammary glands and of other feminine characteristics. Estrogenic steroids have been isolated from urines of pregnant female mammals of many species, including humans, from placental and adrenal tissues, and, unexpectedly, from the testes and urines of stallions.

The corpus luteum, a modification of vertebrate ovarian tissue that forms following ovulation, produces the gestogens, which include progesterone and its derivatives. Gestogens also are secreted by the adrenals and placenta in some species. These compounds, in combination with the estrogens, regulate the metabolism of the uterus to permit implantation and subsequent development of the fertilized ovum in mammals. In birds, estrogens and gestogens stimulate the development of the oviduct and its secretion of albumin. Estrogens and gestogens suppress ovulation (release of the mature egg cell from the ovary); this fact is the basis of action of steroid antifertility drugs (see below Pharmacological actions of steroids: Synthetic estrogens and gestogens: steroid contraceptives). Estrogens and gestogens occur in primitive invertebrates, but their functions in those animals are obscure.

In male vertebrates the androgens--steroids secreted by the testes--maintain spermatogenesis and the tissues of the reproductive tract.

They promote male sexual behaviour and aggressiveness, muscular development, and, in humans, the growth of facial and body hair and deepening of the voice. Testosterone and androstenedione are the principal androgens of the testes. Testosterone is more potent than androstenedione, but in the sexual tissues it appears to be converted to 5 α-dihydrotestosterone, an even more potent androgen.

PERCEPTION OF SEX

It is difficult to assess the degree to which differences related to the sex of the perceiver are biologically based or are the cultural product of traditional differences in sex role. Biological sex and sex role thus far have been hopelessly confounded in experiments with human subjects.

Sex differences in perceiving, whatever their basis, can be illustrated in research on differences in the style with which people perceive. This stylistic difference emerges in extremes of response to context. If a person perceives the world as highly differentiated, he tends to resist contextual influences and is said to be field independent; the person who perceives in an extremely diffuse style, the field-dependent individual, tends to be highly susceptible to contextual effects. Thus, field-independent people are superior in locating a simple visual figure (*e.g.*, a triangle) embedded in a complex pattern; similarly, field-independent subjects can better adjust a rod in a tilted frame to the true vertical when no other visual cues to verticality are present.

Both age and sex are found to be implicated in these differences in perceptual style. Specifically, field dependence declines with increasing age, as does the closely related susceptibility to optical illusions. In North American studies, female subjects tend to be more field dependent than are males, especially after puberty. Perhaps these results are distinctive of cultures in which females are at least implicitly trained to be passive and perceptually diffuse, and in which males are encouraged to assume an active, perceptually articulated stance. This hypothesis has received some support in studies of the parent-child interactions characteristic of the early years of the two types of subject.

ADAPTIVE SIGNIFICANCE OF SEX

When two reproductive cells from somewhat unlike parents come together and fuse, the resulting product of development is never exactly the same as either parent. On the other hand, when new individuals, plant or animal, develop from cuttings, buds, or body fragments, they are exactly like their respective parents, as much alike as identical twins. Any major change in environmental circumstances might exterminate a race since all could be equally affected. When eggs and sperm unite, they initiate development and also establish genetic diversity among the population. This diversity is truly the spice of life and one of the secrets of its success; sex is necessary to its accomplishment.

In each union of egg and sperm, a complete set of chromosomes is contributed by each cell to the nucleus of the fertilized egg. Consequently, every cell in the body inherits the double set of chromosomes and genes derived from the two parental cells. Every time a cell divides, each daughter cell receives exact copies of the original two sets of chromosomes. The process is known as mitosis. Accordingly, any fragment of tissue has the same genetic constitution as the body as a whole and therefore inevitably gives rise to an identical individual if it becomes separated and is able to grow and develop. Only in the case of the tissue that produces the sex cells do cells divide differently, and genetic differences occur as a result.

During the ripening of the sex cells, both male and female, cell divisions (known as meiosis) occur that result in each sperm and egg cell having only a single set of chromosomes. In each case the set of chromosomes is complete--*i.e.*, one chromosome of each kind--but each such set is, in effect, drawn haphazardly from the two sets present in the original cells. In other words, the single set of chromosomes present in the nucleus of any particular sperm or egg, while complete in number and kinds, is a mixture, some chromosomes having come from the set originally contributed by the male parent and some from the female. Each reproductive cell, of either sex, therefore contains a set of chromosomes different in genetic detail from that of every other reproductive cell.

When these in turn combine to form fertilized eggs or fertile seeds, the double set of chromosomes characteristic of tissue cells is re-established, but the genetic constitution of all such cells in the new individual will be the same as that of the fertilized egg--two complete sets of genes, randomly derived from sets contributed by the two different parents. Variation is thus established in two steps.

The first is during the ripening of the sex cells, when each sperm or egg receives a single set of chromosomes of mixed ancestry. None of these cells will have exactly the same combination of genes characteristic of the respective parent. The second step occurs at fertilization, when the pair of already genetically unique sex cells fuse together and their nuclei combine, thus compounding the primary variation.

FRIGIDITY

in psychology, the inability of a woman to attain orgasm during sexual intercourse. In popular, non-medical usage the word has been used traditionally to describe a variety of behaviours, ranging from general coldness of manner or lack of interest in physical affection to aversion to the act of sexual intercourse. Because of the derogatory connotations that have become associated with the term frigidity, it has been replaced in the vocabulary of sex therapists by the general term hypogyneismus, the inability of a woman to obtain sexual satisfaction under otherwise appropriate circumstances.

The lay term frigidity encompasses three distinct problems recognized by sex therapists: inability to experience a sexual response of any kind;

ability to achieve sexual arousal only with great difficulty (hyposexuality); and the inability to achieve orgasm (anorgasmia). Failure of sexual response in females--as in males--may have specific physical sources; such is the case of women who experience vaginal spasms (vaginismus) or pain (dyspareunia) during attempted intercourse. Likewise, female sexual response may be impaired by purely psychological causes, triggered by emotional conflicts outside the sexual relationship or by anxiety and other stresses within the relationship. *See also* sexual dysfunction.

HEREDITY AND SEX LINKAGE

The male of many animals has one chromosome pair, the sex chromosomes, consisting of unequal members called X and Y. At meiosis the X and Y chromosomes first pair, then disjoin and pass to different cells. One-half of the gametes (spermatozoa) formed contain the X and the other half the Y chromosome. The female has two X chromosomes, and all egg cells normally carry a single X. The eggs fertilized by X-bearing spermatozoa give females (XX), and those fertilized by Y-bearing spermatozoa give males (XY).

The genes located in the X chromosomes exhibit what is known as sex-linkage or crisscross inheritance. This is due to a crucial difference between the paired sex chromosomes and the other pairs of chromosomes (called autosomes). The members of the autosome pairs are truly homologous; that is, each member of a pair contains a full complement of the same genes (albeit, perhaps, in different allelic forms). The sex chromosomes, on the other hand, do not constitute a homologous pair, as the X chromosome is much larger and carries far more genes than does the Y. Consequently, many recessive alleles carried on the X chromosome of a male will be expressed just as if they were dominant, for the Y chromosome carries no genes to counteract them.

The classic case of sex-linked inheritance, described by Morgan in 1910, is that of the white eyes in *Drosophila*. White-eyed females crossed to males with the normal red eye colour produce red-eyed daughters and white-eyed sons in the F_1 generation and equal numbers of white-eyed and red-eyed females and males in the F_2 generation. The cross of red-eyed females to white-eyed males gives a different result: both sexes are red eyed in F_1, the females in the F_2 generation are red eyed, half of the males are red eyed, and the other half white eyed.

As interpreted by Morgan, the gene that determines the red or white eyes is borne on the X chromosome, and the allele for red eye is dominant over that for white eye. Since a male receives its single X chromosome from his mother, all sons of white-eyed females also have white eyes. A female inherits one X chromosome from her mother and the other X from her father. Red-eyed females may have genes for red eyes in both of their X chromosomes (homozygotes) or may have one X with the gene for red and the other for white (heterozygotes). In the progeny of heterozygous females one half of the sons will receive the X chromosome with the gene for white and will have white eyes, and the other half will receive the X with the gene for red eyes. The daughters of the heterozygous females crossed with white-eyed males will have either two X chromosomes with the gene for white and hence white eyes or will have one X with white and the other X with the gene for red eyes and will be red-eyed heterozygotes.

In humans, the red-green colour blindness and haemophilia are among many traits showing sex-linked inheritance and consequently are due to genes borne in the X chromosome.

In some animals--birds, butterflies and moths, some fish, and at least some amphibians and reptiles--the chromosomal mechanism of sex determination is a mirror image of that described above. The male has two X chromosomes and the female an X and Y chromosome. Here the spermatozoa all have an X chromosome; the eggs are of two kinds, some with X and others with Y chromosomes, usually in equal numbers. The sex of the offspring is then determined by the egg rather than by the spermatozoon.

Sex-linked inheritance is altered correspondingly. A male homozygous for a sex-linked recessive trait, crossed to a female with the dominant one, gives in the F_1 generation daughters with the recessive trait and heterozygous sons with the corresponding dominant trait. The F_2 generation has recessive and dominant females and males in equal numbers. A male with a dominant trait crossed to a female with a recessive trait gives uniformly dominant F_1 and a segregation in a ratio of 2 dominant males: 1 dominant female: 1 recessive female.

Observations on pedigrees or experimental crosses show that certain traits exhibit sex-linked inheritance; the behaviour of the X chromosomes at meiosis is such that the genes they carry may be expected to exhibit sex-linkage. This evidence still failed to convince some sceptics that the genes for the sex-linked traits were in fact borne in certain chromosomes seen under the microscope. An elegant experimental proof was furnished in 1916 by the U.S. geneticist Calvin Blackman Bridges. As stated above, white-eyed *Drosophila* females crossed to red-eyed males usually produce red-eyed female and white-eyed male progeny. Among thousands of such "regular" offspring there are occasionally found exceptional white-eyed females and red-eyed males.

Bridges constructed the following working hypothesis. Suppose that during meiosis in the female, gametogenesis occasionally goes wrong, and the two X chromosomes fail to disjoin. Exceptional eggs will then be produced carrying two X chromosomes and eggs carrying none. An egg with two X chromosomes coming from a white-eyed female fertilized by a spermatozoon with a Y chromosome will give an exceptional white-eyed female. An egg with no X chromosome fertilized by a spermatozoon with an X chromosome derived from a red-eyed father will yield an exceptional red-eyed male.

This hypothesis can be rigorously tested. The exceptional white-eyed females should have not only the two X chromosomes but also a Y chromosome, which normal females do not have. The exceptional males should, on the other hand, lack a Y chromosome, which normal males do have. Both predictions were verified by examination under a microscope of the chromosomes of exceptional females and males. The hypothesis also predicts that exceptional eggs with two X chromosomes fertilized by X-bearing spermatozoa must give individuals with three X chromosomes; such individuals were later identified by Bridges as poorly viable "super-females." Exceptional eggs with no Xs, fertilized by Y-bearing spermatozoa, will give zygotes without X chromosomes; such zygotes die in early stages of development.

HORMONES IN ANIMALHOOD

Because in most developing animals the reproductive gland is essentially neutral to begin with, there is generally some possibility that agents external to the gland, particularly chemical agents--*i.e.*, hormones--circulating in the blood system, may override the sex-determining influence of the sex chromosomes. In the chick, for example, the sex can be controlled experimentally by such means until about four hours after hatching. If a female chick is injected on hatching with the male sex hormone, testosterone, it will develop into a fully functional cock.

Even when injected at later stages of growth, the male hormone causes extra early growth of the comb, crowing, and aggressive behaviour after being injected in either male or female chicks. Female sex hormones, such as oestrogen, on the other hand, stimulate early growth of the oviduct in the female and feminize the plumage and suppress comb growth when injected in the male.

This susceptibility of the reproductive glands, and sexuality in general, to the influence of sex hormones is particularly acute in mammals, where the egg and embryo, unprotected by any shell, develop in the uterus exposed to various chemicals filtering through from the maternal blood stream. A developing embryo eventually produces its own sex hormones, but they are not manufactured in any quantity until the anatomical sex of the embryo is already well established. One of the curious things about sex hormones, however, is that the reproductive glands are not the only tissues that produce them.

The placenta, through which all exchange between foetus and mother takes place, itself produces tremendous amounts of female sex hormone, together with some male hormone, which are excreted by the mother during pregnancy. This condition is true of humans, as well as of mice and rats. As a rule these hormones are produced too late to do any harm, but not always. The female embryo is fairly immune inasmuch as additional female hormone merely causes a child to be more feminine than usual at an early age. Male embryos, however, may be seriously affected if the female hormone catches them at an early stage. Boy babies may be born that are truly males but under the impact of the feminizing hormone appear superficially to be females and are often raised as such.

As a rule, even when older, they have more or less sterile, undescended testes; an imperfect penis; well-developed breasts; an unbroken voice; and no beard. One in a thousand may be like this and on occasion may have won in women's Olympic competitions. In other cases, those somewhat less severely affected, during adolescence when the hidden testes begin to secrete their own male hormones in abundance, the falsely female characteristics become suppressed, and the voice, beard, breasts, and sexual interest take on the pattern of the male. What were thought to be girls in their youth change into the men they were meant to be upon reaching maturity.

SEASONAL OR PERIODIC SEXUAL CYCLES

In most animals sexual reproduction is seasonal or rhythmical, and so is sexual behaviour, whether in the form of courtship, drive, or other activities that lead to mating. In the marine fireworm of the West Indies, for instance, individuals of both sexes live in crevices on the sea floor but come out to breed where their fertilized eggs can drift and develop in the water above. But they can only find one another by means of the luminescence they themselves produce, which is an eerie light visible only in complete darkness.

Each spring or summer month they emerge and swim to the surface about one-half hour after sunset when all daylight is gone but only before the moon can rise, a situation that confines them to a monthly breeding period of three or four days after the full of the moon. They follow a lunar rhythm. So do the grunion, a common fish along the southern California coast. Here again mating takes place when all is dark and the tide is high. Pairing occurs in the wash of the waves on the sand; fertilized eggs become immediately buried and there develop until the next high spring tides reach and wash the upper level sand nearly two weeks later.

The mysterious biological clocks that apparently all living things possess adjust the rhythms of life to the needs of the particular organism. Some of these timing processes call internal signals on a regular day and night basis; others, on a somewhat longer cycle that keeps pace with the moon rather than the sun; and many, especially in the larger animals, run on a seasonal, or annual, cycle. Many activities are brought into line with the regular changes occurring in the environment. Sex and reproduction, however, are adjusted mainly with regard to two functions; namely, safety while mating, which is therefore commonly in the dark, and the launching of the new generation at a time or season when circumstances are most favourable.

Birds lay eggs, and most mammals deliver their young in early spring, when the months ahead are warm and food is plentiful. Sex for the most part is adjusted to this end. Among the mammals, for example, the period of development within the womb varies greatly, from less than three weeks in the smallest to almost a year in the largest and certain others. Yet with few exceptions, the time for birth is in the spring. The time for mating in most cases is accordingly adjusted to this event: the larger the offspring at birth, the earlier the mating must take place. The horse and the great whales mate in spring and deliver in spring; roe deer mate in summer and deliver in spring; goat and sheep mate in the fall and deliver in spring.

Even the elephant, which has a 22-month pregnancy, delivers in spring but must mate in early summer two years before. In small creatures, however, such as mice, rats, hamsters, and shrews, where the gestation, or pregnancy, period is about three weeks, reproduction is still seasonal, but there is time during the warmer months for several broods to be conceived and raised.

In others, expediency may prevail, and mating may occur at a time to suit the convenience of the pairing animals. The little brown bat, for instance, mates in the fall, and yet ovulation does not take place until winter has passed; the spermatozoa survive the winter in the uterus and fertilize the eggs when they in turn arrive there five or six months later. In some other creatures mating occurs at a convenient time, eggs are fertilized, but development itself is suspended at an early stage for a time so that hatching or birthing, depending on the kind of animal, takes place when circumstances are suitable.

In all of this, the time of the mating season is clearly regulated, both with regard to the physiological condition of the animal and to the environmental conditions. The urge and capacity to mate depends on the ripeness of the gonads, male or female. In most animals, the reproductive glands wax and wane according to the seasons; that is, with an annual rhythm or else with a shorter cycle. Hormones are mainly in control of this rhythm.

Sex hormones, male or female, respectively, are produced by the gonads themselves and cause or maintain their growth and at the same time cause the various secondary sexual characteristics of the male or female individual to become enhanced. Male hormone increases masculinity, even when injected into a female. Female canaries injected with male hormone no longer behave as females and shortly begin to sing loud and long and commence the courtship activities of a male. A hen thus injected grows a larger comb, starts to crow, and begins to strut.

The production of these hormones is in turn controlled by hormones of the pituitary gland. Pituitary hormones stimulate ovarian or testicular tissue, which secretes the sex hormones. The sex hormones not only maintain the growth of the sexual tissues generally but inhibit the secretion of pituitary hormones, so that the process does not get out of hand. The pituitary activity, however, is also influenced by external conditions, particularly by stimuli received indirectly from light.

The annual growth of ovaries or testes that occurs in late winter and early spring in frogs, reptiles, birds, and mammals is initiated by the steadily increasing period of daylight. In response to this changing day length, female frogs are packed with eggs and male frogs are ready to croak by the time the mating period arrives. The large eggs of reptiles and birds are ready to be fertilized, and the males are showing whatever they may have to display at the proper time. In mammals, the female comes into heat, the uterus undergoes the preparatory changes for taking care of fertilized eggs, and the male usually has but one thought in his mind. But as daylight ceases to lengthen, the sexual drive slowly diminishes.

SEX AND THE EFFECTS OF ENVIRONMENT

Sex chromosomes, however, do not determine sex directly but do so through their control of such cell activities as metabolism and hormone production. Their determinative influence, indirect though it is, may be complete. On the other hand, environmental conditions may play the dominating role. In the case of *Bonellia*, a unique kind of marine worm, all eggs develop into small larvae of a sexually indifferent kind.

Those that settle freely on the sea floor grow into comparatively large females, each of which has a long, broad extension, the proboscis, at its front end. Those larvae that happen to settle on the proboscis of a female, however, fail to grow beyond a certain minute size and become dwarf males, permanently attached to the female body. The sex-determining factor appears to be the environmental carbon dioxide tension, which is relatively high at the surface of living tissue.

HUMAN DEVELOPMENT AND SEX DIMORPHISM

The differential effects on the growth of bone, muscle, and fat at puberty increase considerably the difference in body composition between the sexes. Boys have a greater increase not only in stature but especially in breadth of shoulders; girls have a greater relative increase in width of hips. These differences are produced chiefly by the changes that occur during puberty, but other sex differentiations arise before that time. Some, like the external genital difference itself, develop during foetal life. Others develop continuously throughout the whole growth period by a sustained differential growth rate. An example of this is the greater relative length and breadth of the forearm in the male when compared with whole arm length or whole body length.

Part of the sex difference in pelvic shape antedates puberty. Girls at birth already have a wider pelvic outlet. Thus the adaptation for childbearing is present from an early age. The changes at puberty are concerned more with widening the pelvic inlet and broadening the much more noticeable hips.

DIFFERENTIATION OF THE SEXES

Animals and plants, apart from microscopic kinds of life, consist of enormous numbers of cells coordinated in various ways to form a single organism, and each consists of many different kinds of cells specialized for performing different functions. Certain tissues are set aside for the production of sexual reproductive cells, male or female as the case may be. Whether they are testes or ovaries or, as in some animals and plants, both together in the same parental individual, they are typically contained within the body, and therefore the sex cells usually need to be passed to the outside in order to function.

Only in certain lowly creatures such as hydras is there a simpler state, for in hydras the testes and ovaries form in the outermost layer of cells of the slender, tubular body, and the sex cells when ripe burst directly from the simple, bulging gonads into the surrounding water. With few other exceptions, in all other creatures the gonads are part of the internal tissues and some means of exit is necessary. In some, such as most worms, all that is needed are small openings, or precisely placed pores, in the body wall through which sperm or eggs can escape.

In most others, more is needed and a tubular sperm duct or an oviduct leads from each testis or ovary, through which the sex cells pass to the exterior. This is minimal equipment, except where none is needed. The gonad and its duct is accordingly comparable to other glands in the body; that is, the gland is generally a more or less compact mass of cells of a particular, specialized kind, together with a duct for passage of the product of the tissue to the site of action. Gonads secrete--*i.e.*, produce and transmit--sex cells that usually act outside the body.

Differentiation between the sexes exists, therefore, as the primary difference represented by the distinction between eggs and sperm, by differences represented by nature of the reproductive glands and their associated structures, and lastly by differences, if any, between individuals possessing the male and female reproductive tissues, respectively.

Sex cells, sexual organs, other sexual structures, and sexual distinction between individuals constitute a series of evolutionary advances connected with various changes and persisting needs in the general evolution of animals and, to some degree, of plants as well. In other words, no matter how large or complex a creature may become, it still needs to deliver functional sex cells to the exterior. This condition is almost always the case for sperm cells.

Among aquatic animals, particularly marine animals whose external medium, the ocean, is remarkably similar chemically to the internal body fluid medium of all animals, eggs are also in most cases shed to the exterior, where development of the fertilized eggs can proceed readily. Even so, time and place are important. Starfish, sea urchins, and many others, for instance, accumulate mature eggs and sperm in the oviducts and sperm ducts until an appropriate time when all can be shed at once.

When one member of a group of such creatures begins to spawn, chemicals included in the discharge stimulate other members to do the same, so that a mass spawning takes place. One might say that the more they are together the more variable their offspring may be. This situation actually is the crux of the matter for nearly all forms of life, because while it may be possible for a single individual to possess both male and female gonads, producing both sperm and eggs, it remains generally desirable, if not essential, that eggs be fertilized by sperm produced by another individual. Cross-fertilization results in a much greater degree of variability than does self-fertilization.

The existence of two types of individuals, male and female, is the common means of ensuring that cross-fertilization will be accomplished, since then nothing else is possible. Where the sexes are separate, therefore, all that is necessary is that members of the opposite sex get together at a time and place appropriate for the initial development of fertilized eggs.

Typically, spawning of this sort is a communal affair, with many individuals of each sex discharging sex cells into the surrounding water. This process is only suitable, however, when eggs are without tough protective cases or membranes; that is, only when eggs are readily fertilizable for some time after being shed and while drifting in the sea. In this circumstance there is no need for individuals of the opposite sex to mate in pairs, nor is such mating practiced.

HUMAN EMBRYOLOGY AND THE GENITAL SYSTEM

The genital organs begin to develop in the second month, but for a time sex is not grossly distinguishable. Also, a double set of male and female ducts arise, and not until later does the unneeded set decline. Hence, this period is commonly called the indifferent stage.

GONADS

Sex glands develop in a pair of longitudinal ridges located alongside the mesentery, the anchoring fold of membrane to the gut. The primordial sex cells appear first in the cloacal wall, from which they migrate upward in the gut, pass through its mesentery, and finally invade the genital ridges, where they proliferate. The testes are the earliest type of gonad to organize. They begin by developing testis cords and a testis capsule.

The cords radiate from one focal point at the periphery, and thin fibrous partitions segregate groups of the cords within wedge-shaped compartments. These cords do not gain channels and become semen-producing tubules until near the time of puberty. The ovaries organize somewhat tardily by differentiating an outer portion, the cortex, and a central portion, the medulla.

The cortex contains the primordial sex cells; these become surrounded by a layer of ordinary cells, thereby forming primary ovarian follicles. Both the testes and the ovaries undergo relative shifts from their early sites to lower positions in the body. But only the testes make a bodily descent; this is into the scrotum.

SEXUALITY: COMPLEMENTARY MATING TYPES

The complementarity of both male and female sex cells and male and female individuals is a form of division of labour. Male sex cells are usually motile cells capable of swimming through liquid, either freshwater, seawater, or body fluids, and they contribute the male cell nucleus but little else to the fertilization process. The female cell also contributes its nucleus, together with a large mass of cell substance necessary for later growth and development following fertilization. The female cell, however, is without any capacity for independent movement.

In other words, small male cells (sperm cells, spermatozoa, or male gametes) are burdened with the task of reaching a female cell (egg, ovum, or female gamete), which is relatively large and awaits fertilization. A full complement of genes is contributed by both nuclei, representing contributions by both parents, but, apart from the nucleus, only the egg is equipped or prepared to undergo development to form a new organism. A comparable division of labour is seen in the distinction between male and female individuals.

The male possesses testes and whatever accessory structures may be necessary for spawning or delivery of the sperm, and the female possesses ovaries and what may be needed to facilitate shedding the eggs or to nurture developing young. Accordingly there is the basic sex, which depends on the kind of sex gland present, and sexuality, which depends on the different structures, functions, and activities associated with the sex glands.

DIRECT ANIMAL DEVELOPMENT

If an animal after birth or emergence from an egg differs from the adult in comparatively minor details (apart from not having functional sex organs), the development is said to be direct. There is no larval stage and no metamorphosis. Direct development does not mean, however, that no changes occur between birth and adulthood. One very obvious change is the growth of the animal.

The rate of growth--not absolute increase--is highest in the early stages of postembryonic life; subsequently, growth continues to slow, ceasing completely at the attainment of adulthood. The rate of growth is dependent on many factors, both external (feeding, temperature) and internal. Of the internal factors, the most important are hormones, especially the growth hormone produced by the hypophysis. If the growth hormone is produced in insufficient quantities, the result is dwarfism; if it is produced in excessive quantities, the result is gigantism.

In the case of direct development, the most important change is the attainment of sexual maturity, which is achieved in several steps and involves the action of several hormones. The gonad rudiments and rudiments of the supporting parts of the reproductive system remain inactive long after birth. At the approach of adulthood, however, two sets of hormones come into action: hypophyseal hormones stimulate the gonads to activity, and gonadal hormones (produced by the gonads) cause the supporting sex organs and other sex characters to become fully developed.

To become functional, the gonads must be acted upon by secretions from the hypophysis. In immature females the follicle-stimulating hormone, which alone causes the egg follicles and the oocytes to grow, and the luteinizing hormone stimulate the follicle cells to produce the female sex hormone, estrogen, which effects the development of the uterus, the milk glands, and other characteristics of the female sex. In the male, the same hypophyseal hormones are produced, with the result that the testes start to produce sperm and to secrete the male sex hormone, androgen.

It appears that the luteinizing hormone is the more active in the male sex, being able to cause both spermatogenesis and androgen secretion. Androgen, in turn, brings about the development of the penis, the descent of the testicles before birth, the appearance of typical male hair growth, and other secondary sex characteristics.

SOCIAL BEHAVIOUR IN ANIMALS REPRODUCTION

It is noted that sex is a way of combining desirable genes from different lines, genes that otherwise might slowly or never get together. In many lines of animals, parental behaviour is clearly useful in protecting or teaching the young. This normally requires the adult to have fewer young.

The careful parent loses in time and energy and number of offspring but comes to prevail in evolution if it has more descendants than does a careless parent that lets its young die. The careless parent prevails if it can get more young out by caring for each one less; some parasites are careless parents because each of the young needs little care and a large number must be produced to get to an extremely distant host.

SEX-LINKED INHERITANCE

In humans, there are at least 115 genes located on the X chromosome that have no counterpart on the Y chromosome. The traits governed by these genes thus show sex-linked inheritance.

This type of inheritance has certain unique characteristics:

(1) There is no male-to-male transmission.

(2) The carrier female (heterozygote) has a 50 percent chance of passing the gene to each of her children. Sons who receive the gene will be hemizygotes and will manifest the trait. Daughters who receive the gene will be heterozygotes.

(3) The male with the trait will pass the gene on to all of his daughters, who will be heterozygotes.

(4) Most sex-linked traits are recessively inherited; hence heterozygote daughters will usually be carriers who do not display the trait. They will occasionally show mild symptoms of the trait (heterozygous manifestation) due to the phenomenon whereby one of the X chromosomes is inactivated.

Replication of the DNA prior to mitosis occurs with a definite order for each chromosome, one of the two X chromosomes in the female always being the last. This late-replicating X chromosome forms the Barr, or sex chromatin, body, a small, dense-staining structure found at the rim of the nucleus in cells between divisions.

Moreover, one can generalize that in all individuals, including those with abnormal numbers of X chromosomes, in any cell all the X chromosomes except one will produce Barr bodies. Evidence has borne out the essential correctness of a hypothesis advanced by Liane B. Russell (1961) and Mary Lyon (1962) to account for this. The so-called Lyon principle states that only one X chromosome in any cell is fully active, with the other (or others) being genetically inactive and producing a Barr body.

Early in prenatal development, at the stage when the embryo consists of about 1,000 cells, all but one of the X chromosomes in each cell is randomly inactivated; the remaining X chromosome stays active throughout all the succeeding generations of the cell's offspring. As a result, the normal female may be described as a mosaic of X-chromosomal activity, since in some cells her paternally derived X and in others her maternally derived X is genetically active. An exception to this rule is the precursor of the female germ cell (oocyte), in which both X chromosomes remain active.

The Barr-body status of an individual can be determined by the appropriate staining of cells scraped from the inside of the mouth and smeared on a slide. The number of Barr bodies a person has on such cells is one less than the number of X chromosomes in the karyotype. The Lyon principle explains the phenomenon of dosage compensation, the fact that for the great majority of genes on the X chromosome the female has the same amount of their products as the male. Thus, although the gene that codes for anti-haemophilic globulin (AHG) is located on the X chromosome, the normal female with two X chromosomes does not produce twice as much AHG as the normal male.

HAEMOPHILIA A

Haemophilia A is the most widespread form of haemophilia, is a relatively common sex-linked, recessively inherited disease. It results from a mutation of the gene for AHG synthesis. Because of the mutation, the affected male cannot produce AHG and his blood fails to clot properly, causing continuous oozing of blood after minor injuries. Bleeding into joints commonly occurs and may be crippling.

Haemophilia became famous because it affected at least 10 males among the descendants of Queen Victoria, who was a carrier. Therapy consists of avoiding trauma and of administering injections of AHG, which can be made in large amounts through recombinant DNA technology.

In the past the carrier female (heterozygote) has been difficult to diagnose. Although on the average such carriers show 50 percent of the normal amount of AHG, the amount any particular carrier female produces may vary greatly because of the random nature of X-chromosomal inactivation. At the time of X-chromosome inactivation (about the 10th day of embryogenesis), there are relatively few precursors of the liver cells that produce AHG in the adult.

By chance, a majority of these precursor cells--which perhaps number as few as 10--may have the X chromosome with the mutant gene inactivated. (If one tosses a coin 10 times, occasionally "heads" may come up eight times.) In such a case, the carrier female would have an AHG level within normal limits, and it would be hard to know if she were in fact a carrier. On the other hand, a majority of the precursor cells may have the X chromosome with the "wild type" gene inactive. Such a carrier female will have a very low level of AHG and as a result will show symptoms of haemophilia (heterozygote manifestation).

The diagnosis of the carrier state is particularly important for the female who, despite no family history of haemophilia, has borne an affected son. If the woman is a carrier, she has a 50 percent risk of passing the trait on to any future sons. If the woman is not a carrier, her son's haemophilia resulted from a spontaneous mutation occurring in one of the germ cells involved in his conception, and she does not have an increased risk of bearing more affected sons.

The identification of a restriction-fragment-length polymorphism closely linked to the haemophilia A locus has made it feasible to diagnose the carrier females. This fragment is also used with genetic amniocentesis (see below) to determine whether a male foetus at risk has inherited the haemophilia gene.

DUCHENNE'S MUSCULAR DYSTROPHY (DMD)

Duchenne's muscular dystrophy (DMD) is a relatively common sex-linked recessive disease affecting about three males per 10,000 births. The disease is characterized by progressive muscular deterioration and death by 14 to 18 years of age. Diagnosis of the carrier state may be as difficult as in haemophilia, for the same reasons. The gene for DMD has been localized to the short arm of the X chromosome.

A sex-linked, recessively inherited disease known as the fragile-X syndrome can be identified from cells cultured in a medium deficient in folic acid. The diagnosis depends upon the identification of a small break at the tip of the long arm of the X chromosome in these cells. The typical disease, since it is inherited as a sex-linked recessive, occurs as a result of the single "fragile" X of the hemizygous male; the heterozygous female, with only one of her two X chromosomes being fragile, is a carrier.

The phenotype in the affected adult male usually includes mild to moderate mental retardation, a characteristic facial appearance including large ears and a prominent jaw, unusually large testes, and various behavioural manifestations including voluble and inappropriately jocular speech. This condition is probably the second most common predominantly genetic cause of mental retardation, Down's syndrome being more frequent.

UNISEXUALITY

Unisexuality in biology is the condition of an organism or species capable of producing only male or female gametes (sex cells) but never both. A unisexual organism of a bisexual species is one in which the male and female gonads are found in separate individuals. In plants this condition is often called dioecism. A unisexual species is one in which all individuals are of the same sex. Some species of whiptail lizards, for example, are only female. New individuals grow from eggs that develop without fertilization (parthenogenesis).

SEX MATING

Mating between two individuals of the opposite sex becomes necessary when eggs must be fertilized at or before the time the eggs are shed. Whenever eggs have a protective envelope of any kind through which sperm cannot penetrate, fertilization must take place before the envelope is formed. The envelope may at first be a gluey liquid, which covers the egg and solidifies as a tough egg case, as in all crustaceans, insects, and related creatures. It may be a thick membrane of protein deposited around the egg, as in fishes generally; or it may be a material that swells up as a mass of jelly surrounding the eggs after the eggs have been shed, as in frogs and salamanders. And finally, it may be a calcified shell, as in birds and reptiles. In all of these organisms the sperm must reach the egg before the protective substance is added, except in those forms in which a small opening or pore persists in the egg membrane through which sperm can enter.

When and how such eggs need to be fertilized depends on the nature of the protective membranes and the time and place of their formation. The jelly surrounding frog and toad eggs, for instance, swells up immediately after the eggs are shed. Mating and fertilization must take place at the time of spawning. Male frogs mount the back of female frogs and each clasps his mate firmly around the body, which not only helps press the egg mass downward but brings the cloacal opening of male and female close together. Eggs and sperm are shed simultaneously, and the eggs are fertilized as they leave the female body.

Fish eggs are also fertilized as or shortly after they are shed, although fish have no arms and mating generally is usually no more than a coming together of the two sexes side by side, so that simultaneous shedding of sperm and eggs can be accomplished. In other creatures the mating procedure may be much more complicated, depending on various circumstances. Crustaceans such as crabs and lobsters, for example, mate in somewhat the same manner as frogs, with the male holding on to the female by means of claw-like appendages and depositing sperm at the openings of the oviducts, which are typically situated near the middle of the under-surface of the body.

IMPOTENCE

Impotence in general, is the inability of a man to achieve or maintain penile erection and hence the inability to participate fully in sexual intercourse. In its broadest sense the term impotence refers to the inability to become sexually aroused; in this sense it can apply to women as well as to men. In common practice, however, the term has traditionally been used to describe only male sexual dysfunctions. Professional sex therapists, while they identify two distinct dysfunctions as forms of impotence, prefer not to use the term impotence per se. Thus, because of its pejorative connotation in lay usage and because of confusion about its definition, the word impotence has been eliminated from the technical vocabulary.

Traditionally, erectile impotence (the classical definition of impotence) is the failure to achieve penile erection during intercourse. It may have either physical or psychological causes. Alcoholism, endocrine disease, and neurological disorders are typical physical causes. Psychological causes include anxiety or hostility toward the sexual partner and work-related stress or other emotional conflicts outside of the relationship. Erectile impotence occasionally occurs with age and, although attributed by the individual to the aging process itself, it is usually secondary to disorders of aging, such as prostate disease.

In ejaculatory impotence, the male achieves an erection but cannot reach orgasm in the partner's vagina. The erection may be maintained for long periods, even long after the female partner has achieved orgasm. This form of impotence nearly always has an emotional rather than physical cause. *See also* sexual dysfunction.

SEX THERAPY

Sex Therapy is a form of behaviour modification or psychotherapy directed specifically at difficulties in sexual interaction. Many sex therapists use techniques developed in the 1960s by the Americans William Masters and Virginia Johnson to help couples with non-organic problems that affect their sex lives, including premature ejaculation, impotence, and other forms of sexual dysfunction.

In the Masters and Johnson technique, a sex history is first taken and the couple given physical examinations to rule out physical problems. Therapists then employ exercises focusing on the giving and receiving of sensual, but not necessarily sexual, pleasure to help the couple overcome anxieties about sex. Specialized treatments directed against specific sex-related problems are also used during therapy. The therapy process often takes place in an intensive marital workshop lasting for several days.

Although the Masters and Johnson approach involves both members of the couple, sex therapy can take other forms. Co-marital therapy refers to the Masters and Johnson model, in which both members of the couple are treated by a team consisting of one male and one female therapist. The couple approach recognizes that sexual dysfunctions take place in the context of the interaction between two people and are not the exclusive problem of one member of the pair. Individual therapy is employed for those without cooperating partners and may involve the use of a surrogate partner or may focus on exercises that can be practiced by an individual to improve his or her sexual interactions. Group therapy, in which individuals discuss feelings about sex, is also employed for both single-sex and male-female groups.

BISEXUALITY

Bisexuality, in biology, is the condition of an organism capable of producing both male and female gametes (sex cells). In plants and micro-organisms, this is often referred to as monoecism. In multi-cellular animals, bisexuality is usually called hermaphroditism .

HERMAPHRODITISM

Hermaphroditism is the condition of having both male and female reproductive organs. Hermaphroditic plants (most flowering plants) are called monoecious, or bisexual. Hermaphroditic animals, mostly invertebrates such as worms, bryozoans (moss animals), trematodes (flukes), snails, slugs, and barnacles, are usually parasitic, slow-moving, or permanently attached to another animal or plant.

In human beings, hermaphroditism is an extremely rare sex anomaly in which gonads for both sexes are present, in which external genitalia may show traits of both sexes, and in which the chromosomes show male-female mosaicism (where one individual possesses both the male XY and female XX chromosome pairs). Choice of sex must be made at birth, usually on the basis of the condition of the external genitalia (*i.e.*, which sex organs predominate), after which surgery is performed to remove the gonads of the opposite sex. The remaining genitalia are then reconstructed to resemble those of the chosen sex.

Individuals with the external appearance of one sex but the chromosomal constitution and reproductive organs of the opposite sex are examples of pseudo-hermaphroditism.

PSYCHOMOTOR LEARNING IN SEX

Although the assessment of sexual differences in perceptual and reactive abilities is complicated by a number of factors (*e.g.*, age, race, and personality), girls and women tend to be more proficient than boys and men in such psychomotor skills as finger dexterity and inverted-alphabet printing. On the other hand, males generally do better than females at pursuit tracking, repetitive tapping, maze learning, and reaction-time tasks. On rotary pursuit-meter tests, women are not only less accurate but more variable than men of the same age and race.

Although males appear to be superior to females in aptitude and capacity, these advantages disappear when subgroups are carefully matched for initial ability. In contrast, speed scores on discrimination-reaction tests reveal clearly diverging trends for college men and women trained intensively for several days (960 trials). This seems to be a genuine sex difference rather than an element of measurement or selection.

Though both groups were equated for intelligence and had similar error scores, females began to suffer cumulative impairment on the fourth day of practice, whereas males kept improving. Sizable average differences in reaction latency as well as in movement time are characteristic of the sexes on other tasks.

Whereas girls tend to attain their maximum proficiency in speeded tasks earlier in life than boys do, males continue to gain over a longer period and maintain their superiority over females for about half a century of the lifespan. After puberty, boys excel at most athletic skills demanding stamina and strength (*e.g.*, jumping, running, throwing). Thus, female Olympic swimming and track-and-field records are inferior to those of males and are achieved by girls who are noticeably younger than male champions in the same events.

Sex has also been implicated in experiments employing complex coordinators, mirror tracers, and selective mathometers, with boys and men typically surpassing girls and women. Not all psychomotor differences associated with sex are intrinsically biological; unequal opportunities, distinctive social learning, role playing, and other culturally conditioned influences undoubtedly modulate the learning and execution of skills by males and females.

CROSS-FERTILIZATION

Cross-fertilization occurs in individuals and in all animal species in which there are separate male and female individuals. Even among hermaphrodites--*i.e.*, those organisms in which the same individual produces both sperm and eggs--many species possess well-developed mechanisms that ensure cross-fertilization. Moreover, many of the hermaphroditic species that are capable of self-fertilization also have capabilities for cross-fertilization.

There are a number of ways in which the sex cells of two separate individuals can be brought together. In lower plants, such as mosses and liverworts, motile sperm are released from one individual and swim through a film of moisture to the egg-bearing structure of another individual. In higher plants, cross-fertilization is achieved via cross-pollination, when pollen grains (which give rise to sperm) are transferred from the cones or flowers of one plant to egg-bearing cones or flowers of another. Cross-pollination may occur by wind, as in conifers, or via symbiotic relationships with various animals (*e.g.*, bees and certain birds and bats) that carry pollen from plant to plant while feeding on nectar.

Methods of cross-fertilization are equally diverse in animals. Among most species that breed in aquatic habitats, the males and females each shed their sex cells into the water and external fertilization takes place. Among terrestrial breeders, however, fertilization is internal, with the sperm being introduced into the body of the female. Internal fertilization also occurs among some fishes and other aquatic breeders.

By recombining genetic material from two parents, cross-fertilization helps maintain a greater range of variability for natural selection to act upon, thereby increasing a species' capacity to adapt to environmental change.

HUMAN DEVELOPMENT

This refers to the development of the reproductive organs and secondary sex characteristics.

The adolescent spurt in skeletal and muscular dimensions is closely related to the rapid development of the reproductive system that takes place at this time. The acceleration of penis growth begins on average at about age $12^{1}/_{2}$ years, but sometimes as early as $10^{1}/_{2}$ and sometimes as late as $14^{1}/_{2}$. The completion of penis development usually occurs at about age $14^{1}/_{2}$, but in some boys is at $12^{1}/_{2}$ and in others at $16^{1}/_{2}$. There are a few boys, it will be noticed, and who do not begin their spurts in height or penis development until the earliest maturers have entirely completed theirs. At ages 13, 14, and 15 there is an enormous variability among any group of boys, who range all the way from practically complete maturity to absolute preadolescence. The same is true of girls aged 11, 12, and 13.

The psychological and social importance of this difference in the tempo of development, as it has been called, is great, particularly in boys. Boys who are advanced in development are likely to dominate their contemporaries in athletic achievement and sexual interest alike.

Conversely the late developer is the one who all too often loses out in the rough and tumble of the adolescent world, and he may begin to wonder whether he will ever develop his body properly or be as well endowed sexually as those others whom he has seen developing around him. An important part of the educationist's and the doctor's task at this time is to provide information about growth and its variability to preadolescents and adolescents and to give sympathetic support and reassurance to those who need it.

The sequence of events, though not exactly the same for each boy or girl, is much less variable than the age at which the events occur. The first sign of puberty in the boy is usually an acceleration of the growth of the testes and scrotum with reddening and wrinkling of the scrotal skin. Slight growth of pubic hair may begin about the same time but is usually a trifle later. The spurts in height and penis growth begin on average about a year after the first testicular acceleration.

Concomitantly with the growth of the penis, and under the same stimulus, the seminal vesicles, the prostate, and the bulbo-urethral glands, all of which contribute their secretions to the seminal fluid, enlarge and develop. The time of the first ejaculation of seminal fluid is to some extent culturally as well as biologically determined but as a rule is during adolescence and about a year after the beginning of accelerated penis growth.

Axillary (armpit) hair appears on average some two years after the beginning of pubic hair growth; that is, when pubic hair is reaching stage 4. There is enough variability and dissociation in these events, so that a very few children's axillary hair actually appears first. In boys, facial hair begins to grow at about the time that the axillary hair appears. There is a definite order in which the hairs of moustache and beard appear: first at the corners of the upper lip, then over all the upper lip, then at the upper part of the cheeks, in the midline below the lower lip, and, finally, along the sides and lower borders of the chin.

The remainder of the body hair appears from about the time of first axillary hair development until a considerable time after puberty. The ultimate amount of body hair that an individual develops seems to depend largely on heredity, though whether because of the kinds and amounts of hormones secreted or because of variations in the reactivity of the end organs is not known.

Breaking of the voice occurs relatively late in adolescence. The change in pitch accompanies enlargement of the larynx and lengthening of the vocal cords, caused by the action of the male hormone testosterone on the laryngeal cartilages. There is also a change in quality that distinguishes the voice (more particularly the vowel sounds) of both male and female adults from that of children. This is caused by the enlargement of the resonating spaces above the larynx, as a result of the rapid growth of the mouth, nose, and maxilla (upper jaw).

In the skin, particularly of the armpits and the genital and anal regions, the sebaceous and apocrine sweat glands develop rapidly during puberty and give rise to a characteristic odour; the changes occur in both sexes but are more marked in the male. Enlargement of the pores at the root of the nose and the appearance of comedones (blackheads) and acne, while likely to occur in either sex, are considerably more common in adolescent boys than girls, since the underlying skin changes are the result of androgenic (male sex hormone) activity.

During adolescence the male breast undergoes changes, some temporary and some permanent. The diameter of the areola, which is equal in both sexes before puberty, increases considerably, though less than it does in girls. In some boys (between a fifth and a third of most groups studied) there is a distinct enlargement of the breast (sometimes unilaterally) about midway through adolescence. This usually regresses again after about one year.

In girls the start of breast enlargement--the appearance of the "breast bud"--is as a rule the first sign of puberty, though the appearance of pubic hair precedes it in about one-third. The uterus and vagina develop simultaneously with the breast. The labia and clitoris also enlarge. Menarche, the first menstrual period, is a late event in the sequence. Though it marks a definitive and probably mature stage of uterine development, it does not usually signify the attainment of full reproductive function.

The early cycles may be more irregular than later ones and in some girls, but by no means all, are accompanied by discomfort. They are often anovulatory; that is, without the shedding of an egg. Thus there is frequently a period of adolescent sterility lasting a year to 18 months after menarche, but it cannot be relied on in the individual case. Similar considerations may apply to the male, but there is no reliable information about this.

On average, girls grow about six centimetres (about 2.4 inches) more after menarche, though gains of up to twice this amount may occur. The gain is practically independent of whether menarche occurs early or late.

REPRODUCTIVE BEHAVIOUR - DISPLAYS

It has been pointed out that, in general, animals have relatively few displays; in addition, it has been deduced that the relative stability of displays is a dynamic equilibrium--that is, new ones are gained and old ones are lost at about the same frequency. Displays are lost when they no longer convey a selective advantage to the individuals using them; that is, when they are no longer effective in promoting the behaviour that seeks to maximize gene survival in the next generation.

New displays, on the other hand, generally arise by ritualization of previously existing behaviours or functions; that is, when a selective advantage accrues to those individuals who, to convey information, use certain behaviours or functions in a manner that is either partly or totally different from their original purpose. Pheromones, for example, are usually derived from compounds that are natural breakdown products of body metabolism, such as the compounds in urine. Thus, urine, as the precursor of these chemical sex attractants in insects, functions for display purposes, which is far removed from its basic excretory function.

Darwin proposed a theory of sexual selection to account for the presence in animals of displays and functions that apparently were not related to survival. He pointed out that two general concepts were involved. First, the evolution of such characteristics as the larger size of males in many species and the development of horns and antlers in mammals could be accounted for by their usefulness in fights between males for their sexual possession of females. This concept has been termed intra-sexual selection. For such colourful male structures as the plumes of birds of paradise and the tails of peacocks, Darwin suggested that they resulted from the cumulative effects of sexual preference exerted by the females of the species at the time of mating. This second concept has been termed epigamic selection.

A displaying male has been known to convey information about his relative fitness; that is, his ability, with respect to other displaying males, to maximize the survival of his genes into the next generation. Both the brightness of his coloration and the frequency with which he struts say something about the effectiveness of his genes to produce a "healthy" individual. Once this correlation takes place, selection favours those females who are able to choose the "most fit" males. Correspondingly, sexual selection intensifies the signals up to the point at which any further elaboration of those signals would result in a loss of fitness. When selection goes beyond this point, the male, because of his elaborate ornamentation and other displays, is more likely to suffer from predation before he has the opportunity to reproduce.

ARTHROPOD - REPRODUCTIVE SYSTEM AND LIFE CYCLE

With few exceptions, the sexes are separate in arthropods; *i.e.*, there are both male and female individuals. The paired sex organs, or gonads, of each sex are connected directly to ducts that open onto the ventral surface of the trunk, the precise location depending upon the arthropod group.

In arthropods, sperm are commonly transferred to the female within sealed packets known as spermatophores. In this method of transfer the sperm are not diluted by the surrounding medium, in the case of aquatic forms, nor do they suffer from rapid desiccation on land. Among some arachnids, such as scorpions, pseudoscorpions, and some mites, the stalked spermatophore is deposited on the ground.

Either the female is attracted to the spermatophore chemically or the deposition of the spermatophore occurs during the course of a nuptial dance, and the male afterward maneuvers the female into a position in which she can take up the spermatophore within her genital opening.

Centipedes also utilize spermatophores with an accompanying courtship behaviour. Among insects there are some primitive wingless groups, such as collembolans and thysanurans, in which the spermatophore is deposited on the ground, but in most insects the spermatophores are placed directly into the female genital opening by the male during copulation. Many arthropods transfer free sperm rather than spermatophores. These include many crustaceans, millipedes, some insects (such as dipterans and hemipterans), spiders, and some mites.

Arthropod eggs are usually rich in yolk, but in all groups there are species whose eggs have little yolk. Some specialized methods of reproduction found among certain arthropods include the development of unfertilized eggs (parthenogenesis), the birth of living young (viviparity), and the formation of several embryos from a single fertilized egg (polyembryony).

The eggs of many crustaceans hatch into larvae which have fewer segments than the adult. The earliest larval hatching stage is a minute nauplius larva, which possesses only the first three pairs of appendages. Additional segments and appendages then appear at regular intervals with molting. There are several advantages of larval stages in the development of aquatic animals: Currents disperse the larvae, enabling some to settle in different locations from the parents; because many larvae are capable of feeding, less yolk is required in the egg; and, moreover, planktonic larvae do not compete with benthic adults.

In most chelicerates and insects, almost all of the segments are present at hatching, although in insects the body form may differ from that of the adult. Primitive insects, such as collembolans, have the adult form on hatching. Many insects, such as grasshoppers, crickets, and true bugs, hatch as nymphs, which superficially resemble the adult but lack wings. They gradually acquire these adult features during the nymphal instars. Other insects, such as beetles, butterflies, moths, flies, and wasps, hatch as larvae (grubs, caterpillars, maggots) that differ markedly from the adult. The larvae inhabit different environments and eat different foods than the adults. In these insects a pupal stage with metamorphosis bridges the gap between the larva and the adult form.

Myriapods have the general body form of the adult on hatching though they may lack some of the segments. Most millipedes hatch with only seven trunk segments. Some centipedes hatch with all of the adult trunk segments, but others have fewer than the adult.

The young of most arachnids are similar to the adult. The female scorpion gives birth to her young, which immediately climb onto her back. Female wolf spiders also carry their young, and prior to hatching they carry the white egg case attached to the posterior spinnerets. Unlike other arachnids, mites and ticks hatch as six-legged larvae, which acquire the fourth pair of legs at a later molt.

3. HUMAN BEHAVIOUR
SOCIAL AND CULTURAL ASPECTS

The effects of societal value systems on human sexuality are profound. The American anthropologist George P. Murdock summarized the situation, saying:

All societies have faced the problem of reconciling the need of controlling sex with that of giving it adequate expression, and all have solved it by some combination of cultural taboos, permissions, and injunctions. Prohibitory regulations curb the socially more disruptive forms of sexual competition. Permissive regulations allow at least the minimum impulse gratification required for individual well-being. Very commonly, moreover, sex behaviour is specifically enjoined by obligatory regulations where it appears directly to subserve the interests of society.

The historical heritage is, of course, the foundation upon which the current situation rests. Western civilizations are basically Greco-Roman in social organization, philosophy, and law, with a powerful admixture of Judaism and Christianity. This historical mixture contained incompatible elements: individual freedom was cherished, yet there was a great emphasis on law and proper procedure; the pantheism of the Greeks and Romans clashed with Judeo-Christian monotheism; and the sexual permissiveness of Hellenistic times was answered by the anti-sexuality of early Christianity.

In terms of sex, the most important factor was Christianity. While other vital aspects of human life, such as government, property rights, kinship, and economics, were influenced to varying degrees, sexuality was singled out as falling almost entirely within the domain of religion. This development arose from an ascetic concept shared by a number of religions, the concept of the good spiritual world as opposed to the carnal materialistic world, the struggle between the spirit and the flesh. Since sex epitomizes the flesh, it was obviously the enemy of the spirit.

Beginning in the 2nd century, Western Christianity was heavily influenced by this dichotomous philosophy of the Gnostics; sex in any form outside of marriage was an unmitigated evil and, within marriage, an unfortunate necessity for purposes of procreation rather than pleasure. The powerful anti-sexuality of the early Christians (note that neither God nor Christ has a wife and that marriage does not exist in heaven) was in part due to their apocalyptic vision of life: they anticipated that the end of the world and the Last Judgment would soon be upon them. There was no time for a gradual weaning away from the flesh; an immediate and drastic approach was necessary. Indeed, such excessive anti-sexuality developed that the church itself was finally moved to curb some of its more extreme forms.

As it became evident that human existence was going to continue for some unforeseeable length of time and as occasional intelligent theologians made themselves felt, anti-sexuality was ameliorated to some extent but still remained a foundation stone of Christianity for centuries. This attitude was particularly unfortunate for women, to whom most of the sexual guilt was assigned. Women, like the original temptress Eve, continued to attract men to commit sin. They were spiritually weak creatures prone to yield to carnal impulses. This is, of course, a classic example of projecting one's own guilty desires upon someone else.

Ultimately, legal control over sexual behaviour passed from the church to the state, but in most instances the latter simply perpetuated the attitudes of the former. Priests and clergymen frequently continued to exert powerful extralegal control: denunciations from the pulpit can be as effective as statute law in some cases. Although religion has weakened as a social control mechanism, even today liberalization of sex laws and relaxation of censorship have often been successfully opposed by religious leaders. On the whole, however, Christianity has become progressively more permissive, and sexuality has come to be viewed not as sin but as a God-given capacity to be used constructively.

Apart from religion, the state sometimes imposes restrictions for purely secular reasons. The more totalitarian a government, the more likely it is to restrict or direct sexual behaviour. In some instances, this comes about simply as the consequence of a powerful individual (or individuals) being in a position to impose ideas upon the public. In other instances, one cannot escape the impression that sex, being a highly personal and individualistic matter, is recognized as antithetical to the whole idea of strict governmental control and supervision of the individual. This may help explain the rigid censorship exerted by most totalitarian regimes over sexual expression. It is as though such a government, being obsessed with power, cannot tolerate the power the sexual impulse exerts on the population.

PSYCHOLOGICAL EFFECTS OF EARLY CONDITIONING

Physiology sets only very broad limits on human sexuality; most of the enormous variation found among humans must be attributed to the psychological factors of learning and conditioning.

The human infant is born simply with the ability to respond sexually to tactile stimulation. It is only later and gradually that the individual learns or is conditioned to respond to other stimuli, to develop a sexual attraction to males or females or both, to interpret some stimuli as sexual and others as nonsexual, and to control in some measure his or her sexual response. In other words, the general and diffuse sexuality of the infant becomes increasingly elaborated, differentiated, and specific.

The early years of life are, therefore, of paramount importance in the development of what ultimately becomes adult sexual orientation. There appears to be a reasonably fixed sequence of development. Before age five, the child develops a sense of gender identity, thinks of himself or herself as a boy or girl, and begins to relate to others differently according to their gender. Through experience the child learns what behaviour is rewarded and what is punished and what sorts of behaviour are expected of him or her.

Parents, peers, and society in general teach and condition the child about sex not so much by direct informational statements and admonitions as by indirect and often unconscious communication. The child soon learns, for example, that he can touch any part of his body or someone else's body except the anal-genital region. The child rubbing its genitals finds that this quickly attracts adult attention and admonishment or that adults will divert him or her from this activity. It becomes clear that there is something peculiar and taboo about this area of the body.

This "genital taboo" is reinforced by the great concern over the child's excretory behaviour: bladder and bowel control is praised; loss of control is met by disappointment, chiding, and expressions of disgust. Obviously, the anal-genital area is not only a taboo area but a very important one as well. It is almost inevitable that the genitalia become associated with anxiety and shame. It is noteworthy that this attitude finds expression in the language of Western civilizations, as in "privates" (something to be kept hidden) and the German word for the genitals, *Scham* ("shame").

While all children in Western civilizations experience this antisexual teaching and conditioning, a few have, in addition, atypical sexual experiences, such as witnessing or hearing sexual intercourse or having sexual contact with an older person. The effects of such atypical experiences depend upon how the child interprets them and upon the reaction of adults if the experience comes to their attention.

Seeing parental coitus is harmless if the child interprets it as playful wrestling but harmful if he considers it as hostile, assaultive behaviour. Similarly, an experience with an adult may seem merely a curious and pointless game, or it may be a hideous trauma leaving lifelong psychic scars. In many cases the reaction of parents and society determines the child's interpretation of the event. What would have been a trivial and soon-forgotten act becomes traumatic if the mother cries, the father rages, and the police interrogate the child.

Some atypical developments occur through association during the formative years. A child may associate clothing, especially underclothing, stockings, and shoes with gender and sex and thereby establish the basis for later fetishism or transvestism. Others, having been spanked or otherwise punished for self-masturbation or childhood sex play, form an association between punishment, pain, and sex that could escalate later into sadism or masochism. It is not known why some children form such associations whereas others with apparently similar experience do not.

Around the age of puberty, parents and society, who more often than not refuse to recognize that children have sexual responses and capabilities, finally face the inescapable reality and consequently begin inculcating children with their attitudes and standards regarding sex. This campaign by adults is almost wholly negative--the child is told what not to do. While dating may be encouraged, no form of sexual activity is advocated or held up as model behaviour. The message usually is "be popular" (*i.e.*, sexually attractive), but abstain from sexual activity.

This anti-sexualism is particularly intense regarding young females and is reinforced by reference to pregnancy, venereal disease, and, most importantly, social disgrace. To this list religious families add the concept of the sinfulness of premarital sexual expression. With young males the double standard of morality still prevails. The youth receives a double message, "don't do it, but we expect that you will." No such loophole in the prohibitions is offered young girls. Meanwhile, the young male's peer group is exerting a pro-sexual influence, and his social status is enhanced by his sexual exploits or by exaggerated reports thereof.

As a result of this double standard of sexual morality, the relationship between young males and females often becomes a ritualized contest, the male attempting to escalate the sexual activity and the female resisting his efforts. Instead of mutuality and respect, one often has a struggle in which the female is viewed as a reluctant sexual object to be exploited, and the male is viewed as a seducer and aggressor who must succeed in order to maintain his self-image and his status with his peers.

This sort of pathological relationship causes a lasting attitude on the part of females: men are not to be trusted; they are interested only in sex; a girl dare not smile or be friendly lest males interpret it as a sign of sexual availability, and so forth. Such an aura of suspicion, hostility, and anxiety is scarcely conducive to the development of warm, trusting relationships between males and females. Fortunately, love or infatuation usually overcomes this negativism with regard to particular males, but the average female still maintains a defensive and skeptical attitude toward men.

Western society is replete with attitudes that impede the development of a healthy attitude toward sex. The free abandon so necessary to a full sexual relationship is, in the eyes of many, an unseemly loss of self-control, and self-control is something one is urged to maintain from infancy onward. Panting, sweating, and involuntary vocalization are incompatible with the image of dignity. Worse yet is any substance once it has left the body: it immediately becomes unclean. The male and female genital fluids are generally regarded with disgust--they are not only excretions but sexual excretions. Here again, societal concern over excretion is involved, for sexual organs are also urinary passages and are in close proximity to the "dirtiest" of all places--the anus. Lastly, many individuals in society regard menstrual fluid with disgust and abstain from sexual intercourse during the four to six days of flow. This attitude is formalized in Judaism, in which menstruating females are specifically labelled as ritually unclean.

In view of all these factors working against a healthy, rational attitude toward sex and in view of the inevitable disappointments, exploitations, and rejections that are involved in human relationships, one might wonder how anyone could reach adulthood without being seriously maladjusted. The sexual impulse, however, is sufficiently strong and persistent and repeated sexual activity gradually erodes the inhibitions and any sense of guilt or shame.

Further, all humans have a deep need to be esteemed, wanted, and loved. Sexual activity with another is seen as proof that one is attractive, desired, valued, and possibly loved--a proof very necessary to self-esteem and happiness. Hence, even among the very inhibited or those with weak sex drive, there is this powerful motivation to engage in socio-sexual activity.

Most persons ultimately achieve at least a tolerable sexual adjustment. Some unfortunates, nevertheless, remain permanently handicapped, and very few completely escape the effects of society's antisexual conditioning. While certain inhibitions and restraints are socially and psychologically useful--such as deferring gratification until circumstances are appropriate and modifying behaviour out of regard for the feelings of others--most people labour under an additional burden of useless and deleterious attitudes and restrictions.

NERVOUS SYSTEM FACTORS

The nervous system consists of the central nervous system and the peripheral nervous system. The brain and spinal cord constitute the central system, while the peripheral system is composed of the:

(1) cerebrospinal nerves that go to the spinal cord (afferent nerves), transmitting sensory stimuli and those that come from the cord (efferent nerves) transmitting impulses to activate muscles, and

(2) autonomic system, the primary function of which is the regulation and maintenance of the body processes necessary to life, such as heart rate, breathing, digestion, and temperature control. Sexual response involves the entire nervous system.

The autonomic system controls the involuntary responses; the afferent cerebrospinal nerves carry the sensory messages to the brain; the efferent cerebrospinal nerves carry commands from the brain to the muscles; and the spinal cord serves as a great transmission cable. The brain itself is the coordinating and controlling centre, interpreting what sensations are to be perceived as sexual and issuing appropriate "orders" to the rest of the nervous system.

The parts of the brain thought to be most concerned with sexual response are the hypothalamus and the limbic system, but no specialized "sex centre" has been located in the human brain. Animal experiments indicate that each individual has coded in its brain two sexual response patterns, one for mounting (masculine) behaviour and one for mounted (feminine) behaviour. The mounting pattern can be elicited or intensified by male sex hormone and the mounted pattern by female sex hormone. Normally, one response pattern is dominant and the other latent but capable of being called into action when suitable circumstances occur. The degree to which such inherent patterning exists in humans is unknown.

While the brain is normally in charge, there is some reflex (*i.e.*, not brain-controlled) sexual response. Stimulation of the genital and perineal area can cause the "genital reflex": erection and ejaculation in the male, vaginal changes and lubrication in the female. This reflex is mediated by the lower spinal cord, and the brain need not be involved. Of course, the brain can override and suppress such reflex activity--as it does when an individual decides that a sexual response is socially inappropriate.

SEXUALLY TRANSMITTED DISEASES

Infections transmitted primarily by sexual contact are referred to as sexually transmitted diseases (STDs). Caused by a variety of microbial agents that thrive in warm, moist environments such as the mucous membranes of the vagina, urethra, anus, and mouth, STDs are diagnosed most frequently in individuals who engage in sexual activity with many partners.

In the past, a disease transmitted sexually was more commonly called a venereal disease, or VD, and was applied to only a few infections such as gonorrhea and syphilis. Actually more than 20 STDs have been identified, and infections caused by *Chlamydia trachomatis*, herpes simplex virus, and human papillomavirus, although underreported, are believed to be more prevalent than gonorrhea in the United States. Although the incidence of some STDs has reached epidemic proportions, it was not until the advent of the acquired immunodeficiency syndrome (AIDS) that the need to restrain the transmission of these diseases gained serious attention.

AIDS is a deadly disease for which there is no known cure. This fact has made prevention of the spread of HIV (see below) infection a top priority of the health-care community, with education concerning safer sexual practices at the fore. The "safe sex" strategy, which includes encouraging the use of condoms or the practice of abstinence, has been introduced to prevent the spread not only of AIDS but of all STDs. Stemming the transmission of disease rather than relying on treatment, which in the case of AIDS does not even exist, is the basic tenet of the safe-sex doctrine.

Preventing the transmission of STDs is also important because many of these diseases do not produce initial symptoms of any significance. Thus, they often go untreated, increasing their spread and the incidence of serious complications; untreated chlamydial infections in women are the primary preventable cause of female sterility.

IMPACT OF PSYCHOANALYSIS

There is little doubt that psychoanalysis had a profound influence on personality theory during the 20th century. It turned attention from mere description of types of people to an interest in how people become what they are. Psychoanalytic theory emphasizes that the human organism is constantly, though slowly, changing through perpetual interactions, and that, therefore, the human personality can be conceived of as a locus of change with fragile and indefinite boundaries. It suggests that research should focus not only on studies of traits, attitudes, and motives but also on studies that reflect the psychoanalytic view that personality never ceases to develop and that even the rate of personality modification changes during the course of a life. Although the theory holds that conflict and such basic drives as sex and aggression figure prominently in personality development and functioning, their presence may be neither recognizable nor comprehensible to persons untrained to look for those motives.

However, personality characteristics are relatively stable over time and across situations, so that a person remains recognizable despite change. Another feature of psychoanalytic theory is the insistence that personality is affected by both biological and psychosocial forces that operate principally within the family, with the major foundations being laid early in life.

The data on which psychoanalytic theory rests came from the psychoanalysts' consulting rooms, where patients in conflict told their life stories to their analysts. No provision is made in that setting for experimental manipulation, for independent observation, or for testing the generality of the formulations. As a consequence, although much of the theory has found its way into accepted doctrine, psychoanalysis cannot claim a body of experimentally tested evidence. Nevertheless, psychoanalytic theory provides at least a preliminary framework for much of personality research involving motives and development.

PSYCHOSEXUAL DYSFUNCTION

Sexual dysfunction IS the inability of a person to experience sexual arousal or to achieve sexual satisfaction under appropriate circumstances, as a result of either physical disorder or, more commonly, psychological problems. The most common forms of sexual dysfunction have traditionally been classified as impotence (inability of a man to achieve or maintain penile erection) and frigidity (inability of a woman to achieve arousal or orgasm during sexual intercourse).

Because these terms--impotence and frigidity--have developed pejorative and misleading connotations, they are no longer used as scientific classifications, having been superseded by more specific terms; however, both terms remain in common usage, with a variety of meanings and associations (*see* frigidity; impotence).

Sexual dysfunctions recognized by professional therapists include hyposexuality (or inhibited sexual excitement), in which sexual arousal can be achieved only with great difficulty; anorgasmia, in which a woman has a recurrent and persistent inability to achieve orgasm despite normal sexual stimulation; vaginismus, in which the woman's vaginal muscles contract strongly during intercourse, making coitus difficult or impossible; dyspareunia, in which a woman experiences significant pain during attempts at intercourse; erectile impotence, in which a man cannot sustain an erection; ejaculatory impotence (or inhibited male orgasm), in which a man cannot achieve orgasm in the woman's vagina, although he can sustain an erection and may reach orgasm by other methods; and premature ejaculation, in which the man ejaculates before or immediately after entering the vagina.

In most cases, each of these dysfunctions reflects the individual's anxiety or other negative feelings about the sex act or partner, although emotional conflicts outside the sexual relationship itself can also produce failures of sexual function. Appropriate sex therapy, designed to help the individual relax in his or her sexual role, can often overcome the anxiety and eliminate the dysfunction, although the success of such therapy varies markedly among the various dysfunctions and among individual patients. When a specific physical condition predisposes to the dysfunction, it must be treated medically; alcoholism and endocrine or neurological disorders are among the common physical causes of sexual dysfunction. Sexual dysfunctions that are secondary to more severe psychological or personality disorders may require specific psychotherapy.

GENETIC AND CONGENITAL ABNORMALITIES

In the male, congenital anomalies of the prostate and seminal vesicles are rare; they consist of absence, hypoplasia (underdevelopment), or the presence of fluid- or semisolid-filled sacs, called cysts. Cysts of the prostatic utricle (the uterine remnant found in the male) are often found in association with advanced stages of hypospadias (a defect in the urethra, see below) and pseudohermaphroditism (in which sex glands are present but bodily appearance is ambiguous as to sex; *i.e.,* the secondary sexual characteristics are underdeveloped). Cysts may also cause urinary obstructive symptoms through local pressure on the bladder neck.

Severe anomalies of the penis are rare and are generally associated with urinary or other systemic defects that are incompatible with life. Anomalies are those of absence, transposition, torsion (twisting), and reduplication of the penis. An abnormally large penis frequently is present in boys affected by precocious puberty, in congenital imbeciles, in dwarfs, in men with overactive pituitaries, and in persons affected by adrenal tumours. A small penis is seen in infantilism and in underdevelopment of the genitals, or under-secretion of the pituitary or pineal gland, and failure of development of the corpora cavernosa.

The only anomaly of the foreskin of grave concern is congenital phimosis, characterized by a contracture of the foreskin, or prepuce, sufficient to prevent its retraction over the glans; the preputial opening may be pinhole in size and may impede the flow of urine. The condition is easily remedied by circumcision, a permanent cure.

There is a considerable variety of urethral anomalies. Stenosis (contracture) of the external opening (meatus) is the most common, but congenital stricture of the urethra occasionally occurs at other points. Valves (or flaps) across the anterior or posterior part of the urethra may cause congenital urethral obstruction in boys. Posterior urethral valves are more common than anterior valves and consist of deep folds of mucous membrane, often paper-thin and usually attached at one end to the verumontanum, a small prominence in the back wall of that part of the urethra that is surrounded by the prostate gland. If too tight, the valves may obstruct the urethra and destroy the kidneys.

There are various defects associated with incomplete closure of the urethra. One of the commonest is hypospadias, in which the underside (ventral side) of the urethral canal is open for a distance at its outer end. Frequently the hypospadiac meatus is narrowed, and the penis also has a downward (ventral) curvature beyond the meatus.

The posterior part of the urethra is never involved; therefore, the muscle that closes the urethra, the sphincter, functions normally, and urinary control exists. Although the condition occurs in both sexes, it is seen predominantly in the male. There is a high incidence of partial or complete failure of the testes to develop, cryptorchism (failure of the testes to descend into the scrotum), and small external and internal genitalia; variable male-female admixtures may be associated with this deficiency. Epispadias, an opening in the upper (dorsal) side of the penis, is considerably less common than hypospadias.

Dorsal curvature may also be present, but the disabling aspect is that the defect usually extends through the urinary sphincter and causes urinary incontinence. Other less common urethral anomalies include complete absence of the urethra, double urethra, urethra fistula (an opening in the urethra), urethra-rectal fistula (an opening between the urethra and the rectum), and urethral diverticulum (a pouch in the wall of the urethra). Most of the above conditions are correctable by surgery.

ANARCHISM

Anarchism (absence of one or both testes) is rare; it may be associated with the absence of various other structures of the spermatic tract. Generally, if one testis is absent, the other is found to be within the abdomen rather than in the scrotum. Congenitally small testes may be a primary disorder or may occur because of under-activity of the pituitary. In both disorders, there is a lack of development of secondary sexual characteristics and some deficiency in libido and potency.

Supernumerary testicles are extremely rare; when present, one or more of the supernumerary testicles usually shows some disorder such as torsion of the spermatic cord. Synorchism, the fusion of the two testicles into one mass, may occur within the scrotum or in the abdomen. Cryptorchism is the term applied to all forms of imperfectly descended testes, the commonest anomaly of the spermatic tract. The condition is often bilateral, and in the unilateral cases there is no preponderance between the left or right side. Hormonal treatment may be useful in correcting the condition, but usually surgery is necessary for correction.

PERSONALITY

The dramatic changes that characterize puberty present the adolescent with serious psychosocial challenges. A person who has lived for 12 years has developed a certain sense of self as well as of self-capacity. In adolescence, however, this knowledge of self is challenged. As has been discussed, the rather sudden bodily changes in this period are accompanied by equally dramatic changes in thoughts and feelings.

Thus, not all the assumptions adolescents held about the self in earlier stages may still be relevant to the new individuals they find themselves to be. Because a coherent sense of self is necessary for functioning productively in society, adolescents ask a crucial psychosocial question: Who am I?

At precisely the time that adolescents feel unsure about who they are, society begins to ask them related questions. For instance, adolescents are expected to make the first steps toward career objectives. Society asks adolescents, then, what roles they will play as adults--that is, what socially prescribed set of behaviours they will choose to adopt. Thus, a key aspect of this adolescent dilemma is that of finding a role, which is generally taken to be the outward expression of identity.

The emotional upheaval provoked by this mandate is called the identity crisis. In order to resolve this crisis and achieve a sense of identity, it is necessary to synthesize psychological development and societal directives. The adolescent must find an orientation to life that not only fulfils the attributes of the self but at the same time is consistent with what society expects; that is, a role cannot be self-destructive (*e.g.*, sustained fasting) or socially disapproved (*e.g.*, criminal behaviour). In the search for an identity, the adolescent must discover what he believes in and what his attitudes and ideals are, for commitment to a role entails, to a greater or lesser degree, commitment to a set of values.

If the adolescent fails to resolve the identity crisis by the time of entry into adulthood, he will feel a sense of role confusion or identity diffusion. Some young adults waver between roles in a kind of prolonged "moratorium," or period of avoiding commitment. Others seem to avoid the crisis altogether and settle easily on an available, socially approved identity. Still others resolve their crises by adopting an available but socially disapproved role or ideology. This latter option is called negative identity formation and is often associated with delinquent behaviour. Resolution of the adolescent identity crisis has a profound influence on development during later adulthood.

All societies traditionally prescribe stereotyped roles to each sex. These roles have adaptive significance; that is, they allow society to maintain and perpetuate itself. From this reasoning, it follows that differences in sex-role behaviour, at least initially, arose from the different tasks males and females performed for survival--especially those tasks centred on reproduction. Differing biologies exert differing pressures on psychosocial development; however, these pressures do not occur independently of the demands of cultural and historical milieus.

The biological basis of one's psychosocial functioning is believed to relate to adaptive orientations for survival. Many differences exist between males and females, but the nature of individual differences between the sexes is dependent on interactions among biological, psychological, socio-cultural, and historical influences.

CULTURE AND PERSONALITY

Since the infant of the human species enters the world cultureless, his behaviour--his attitudes, values, ideals, and beliefs, as well as his overt motor activity--is powerfully influenced by the culture that surrounds him on all sides. It is almost impossible to exaggerate the power and influence of culture upon the human animal. It is powerful enough to hold the sex urge in check and achieve premarital chastity and even voluntary vows of celibacy for life. It can cause a person to die of hunger, though nourishment is available, because some foods are branded unclean by the culture. And it can cause a person to disembowel or shoot himself to wipe out a stain of dishonour.

Culture is stronger than life and stronger than death. Among subhuman animals, death is merely the cessation of the vital processes of metabolism, respiration, and so on. In the human species, however, death is also a concept; only man knows death. But culture triumphs over death and offers man eternal life. Thus, culture may deny satisfactions on the one hand while it fulfils desires on the other.

The predominant emphasis, perhaps, in studies of culture and personality has been the inquiry into the process by which the individual personality is formed as it develops under the influence of its cultural milieu. But the individual biologic organism is itself a significant determinant in the development of personality. The mature personality is, therefore, a function of both biologic and cultural factors, and it is virtually impossible to distinguish these factors from each other and to evaluate the magnitude of each in particular cases.

If the cultural factor were a constant, personality would vary with the variations of the neurosensory-glandular-muscular structure of the individual. But there are no tests that can indicate, for example, precisely how much of the taxicab driver's ability to make change is due to innate endowment and how much to cultural experience.

Therefore, the student of culture and personality is driven to work with "modal personalities," that is, the personality of the typical Crow Indian or the typical Frenchman insofar as this can be determined. But it is of interest, theoretically at least, to note that even if both factors, the biologic and the cultural, were constant--which they never are in actuality--variations of personality would still be possible. Within the confines of these two constants, individuals might undergo a number of profound experiences in different chronological permutations.

For example, two young women might have the same experiences of:

(1) having a baby,

(2) graduating from college, and

(3) getting married. But the effect of sequence (1), (2), (3) upon personality development would be quite different than that of sequence (2), (3), (1).

MARITAL CUSTOMS AND LAWS

Some form of marriage has been found to exist in all human societies, past and present. Its importance can be seen in the elaborate and complex laws and rituals surrounding it. Although these laws and rituals are as varied and numerous as human social and cultural organizations, some universals do apply.

The main legal function of marriage is to ensure the rights and define the relationships of the children within a community. Marriage universally confers a legitimate status on the offspring, which entitles him or her to the various privileges set down by the traditions of that community, including the right of inheritance. It also establishes the permissible social relations allowed to the offspring, including the acceptable selection of future spouses.

Until modern times, marriage was rarely a matter of free choice. In Western civilization, love between spouses has come to be associated with marriage. However, romantic love has not been a primary motive for matrimony in most eras, and the person whom it is considered permissible to marry has historically been carefully regulated by most societies.

Endogamy, the practice of marrying someone from within one's own tribe or group, is the oldest social regulation of marriage. When the forms of communication with outside groups are limited, endogamous marriage is a natural consequence. Cultural pressures to marry within one's social, economic, and ethnic group are still very strongly enforced today in some societies.

Exogamy, the practice of marrying outside the group, is found in societies in which kinship relations are the most complex, thus barring from marriage large groups who may trace their lineage to a common ancestor.

In societies in which the large, or extended, family remains the basic unit, marriages are usually arranged by the family. The assumption is that love between the partners comes after marriage, and much thought is given to the socioeconomic advantages accruing to the larger family from the match. By contrast, in societies in which the small, or nuclear, family predominates, young adults usually choose their own mates. It is assumed that love precedes (and determines) marriage, and less thought is normally given to the socioeconomic aspects of the match.

In societies with arranged marriages, the almost universal custom is that someone acts as an intermediary, or matchmaker. This person's chief responsibility is to arrange a marriage that will be satisfactory to the two families represented. Some form of dowry or bridewealth is almost always exchanged in societies that use arranged marriages.

In societies in which individuals choose their own mates, dating is the most typical way for people to meet and become acquainted with prospective partners of the opposite sex. Successful dating may result in courtship, which then usually leads to marriage.

MARITAL ROLES

Marriage is important as the accepted institution for the expression of adult sexuality. A mutually satisfying sex life is important to both men and women, although social scientists point out that marital roles involve much more than this. Romantic love is only one of the reasons people marry. Social and economic security, and indeed social pressures, can be equally important.

Relations between the sexes are to a large extent culturally as well as biologically determined. The image of the "macho" male is well-known and attributed commonly to Mediterranean and Latin-American cultures. In working-class British culture, too, tenderness in a sexual relationship has been traditionally regarded as unmanly. The public image that such men wish to project is based on sexual prowess rather than on emotional intimacy. This image may even be retained after marriage if manliness is defined by how completely a man can rule his household.

In the past, women frequently took their social status from their husbands, but by the late 20th century there was an increasing tendency for women to be regarded as equal partners in marriage. The traditional norm, where women remained at home and men went out to work, has changed rapidly. As women gain status from their own occupations outside the home, they are beginning to achieve equality with men. Women's traditional sphere of influence has been the home, however, and in cultures as diverse as the Khoikhoin (Hottentots) of southern Africa and sections of the working-class population of modern Britain, women's economic authority in the home remains paramount. Even today it is not uncommon for the British husband to depend on his wife to give him spending money, even though it may originate in his wages.

In a study of the family in a low-income area of London, British sociologists Michael Young and Peter Willmott found that what had previously been regarded as the typical late 19th-century family had survived into the 1950s. This type of family was centred on the economic separation of the roles of husband and wife, sometimes with both partners working and frequently with the wife sharing domestic tasks with female relatives who lived nearby. Young married women, for example, received help from their mothers in shopping, household chores, and babysitting. In further studies made in the 1970s, however, Young and Willmott documented changes toward what they called the "symmetrical family," in which kin networks had ceased to be as extensive as in the past and husbands and wives shared domestic tasks between them. Social activities, too, had become more couple-centred, as in many cases men stayed home, perhaps to watch television, rather than to socialize with their male friends. In short, at least in London, there was a development of working-class marital roles toward a pattern similar to that found in most middle-class households. Indications are that the trend is a widespread one.

PEER SOCIALIZATION

During the first two years of life, infants do not spontaneously seek out other children for interaction or for pleasure. Although six-month-old infants may look at and vocalize to other infants, they do not initiate reciprocal social play with them. However, between two and five years of age, children's interactions with each other become more sustained, social, and complex. Solitary or parallel play is dominant among three-year-olds, but this strategy shifts to group play by five years.

PROBLEMS IN DEVELOPMENT

An estimated 6-10 percent of all children develop serious emotional or personality problems at some point. These problems tend to fall into two groups: those characterized by symptoms of extreme anxiety, withdrawal, and fearfulness, on the one hand, and by disobedience, aggression, and destruction of property on the other. The former set is called internalizing; the latter is termed externalizing. As indicated earlier, some fearful, timid, socially withdrawn children inherit a temperamental predisposition to develop this form of behaviour; other children, however, acquire it as a result of a stressful upbringing, experiences, or social circumstances.

Sex-linked differences in aggression are evident from about two or three years of age, with boys being more aggressive than girls. Although young children sometimes fight and quarrel, usually over possessions, such behaviour is generally not a serious problem in the first three or four years of life. Aggressive behaviour can become a serious problem in older children, however, and by seven years of age a small proportion of boys do display an extreme and consistent tendency to be aggressive with others. Children who are highly aggressive by age seven or eight tend to remain so later in life; these children are three times more likely to have police records as adults than are other children. By age 30 significantly more members of this group had been convicted of criminal behaviour, were aggressive with their spouses, and abused or severely punished their own children.

Although biological factors can play a role in producing extreme aggression, the role of the child's social environment is critical. Parents' use of extreme levels of physical punishment, imposed inconsistently, is associated with high levels of aggression in children, as are extreme levels of parental permissiveness toward a child's own aggressive acts. Psychologists frequently help parents deal with aggressive children by teaching them to observe what they do and to enforce rules consistently with their children. Parents can thereby learn effective but nonpunitive ways of controlling their aggressive children.

Although precise information is difficult to obtain, it is estimated that each year about one million children in the United States are abused by their parents or other adults. Child abuse is more common in economically disadvantaged families than in affluent ones but occurs in all social classes, races, and ethnic groups. The abuse of children is often part of a pattern of family violence that is transmitted from parent to child for generations. Children who were abused as infants tend to show much more avoidant, resistant, and noncompliant behaviour than do other children.

4. SEX AND DEVELOPMENTAL PSYCHOLOGY

BABY AND CHILDHOOD SEXUALITY

Although this brief account of baby and childhood sexuality can be verified by the average observant parent, it is not universally accepted. Some people find it hard to believe that events in childhood can exert such profound effect on such matters as the ability to enjoy sex later in life.

If it is accepted that infant and childhood sexuality and the way that is handled are the foundation for what comes in adulthood, then its enormous importance can be readily appreciated. To argue that childhood experiences have no bearing on events in later life is contrary to all the available evidence and to common sense. After all, we happily accept such reasoning on non-sexual matters.

A more subtle and difficult criticism arises in the question of why children who are treated in virtually the same way with regard to sexual and emotional matters display totally different sexualities and sexual problems in adulthood. The answer probably lies in the fact that no two people can really be subject to exactly the same influences, and therefore any two people will respond differently to similar experiences.

How secure children feel in their place in the family also affects their vulnerability to experiences.

Also, the child's own perceptions of what is happening may be different from those of a sibling who is going through the same experience. For these and other reasons, the long term consequences of a similar upbringing can vary enormously.

Parents too are not static personalities – they change as the years pass and they react differently to, and therefore have different influence on, each of the children.

All this makes the study of childhood sexuality a minefield, but an understanding of the processes dealt with can put problems into perspective. We are a product of our yesterdays as well as of our genetic blueprint.

EARLY ADOLESCENCE

Early adolescence is the stage at which girls and boys learn to accept their body changes and emerging sexuality as the start of their progress from childhood to adulthood. Although it is a time of considerable change for both sexes, boys, in general, face a less complicated situation than girls. This parallels the greater complexity the girls experience in early childhood when, unlike boys, they have to switch affection from mother to father. In general, early adolescence is not a particularly stressful time for boys but it can disorganise and distress a girl.

MID-AND LATE ADOLESCENCE

As in several other periods of psycho-sexual development, the girl has more complicated tasks to fulfil in late adolescence. In both sexes the stage is a heterosexual one although by far the most frequent form of sexual expression is masturbation. The sexualised love withdrawn from the opposite-sex parent is invested in the self and this makes late adolescence an important milestone in learning to love.

Late adolescence is also a time of great change, particularly in relationships, and young adulthood is the period of stabilisation which follows it. Romanticism is still rife in late adolescence but it is to be hoped that it is tempered a little by reason in early adulthood. Equally, it is to be hoped that it is never lost

SEXUAL ATTRACTION

Emotional and sexual attractions are complex issues of which we know a few dimensions. Variations in taste between individuals ensure that almost any man or woman will be attractive to someone of the opposite sex. Hairy women, for example, often believe themselves to be unattractive but some men prefer them. Physical differences and even disabilities can be attractive to others. Physical attractiveness is the most important factor to young adolescents and to adults looking for brief affairs. Distinguishing emotional attraction from the physical is mainly an adult skill.

How attractive one feels depends enormously on how one feels generally. A recent good experience, such as a successful flirtation, can increase one's sense of attractiveness. For women particularly, how attractive they feel greatly influences how attractive they are. A good morale is vital to one's sense of attractiveness.

SEXUAL ENCOUNTERS

Sexual experience is gradually acquired in adolescence and very few social factors influence the situation. Society may find certain aspects of teenage sexual activity hard to accept, but it is a reality. Common sense suggests that it makes sense to accept the reality and to do our best to ensure that teenagers come to the least possible physical, emotional and spiritual harm. Total repression of emerging teenage sexuality is a forlorn hope and has negative side-effects both at the time and later.

A goodly proportion of teenage and late-adolescent boys who have had intercourse did not much enjoy the experience. Girls still exaggerate their romantic and loving feelings to justify their sexual adventures. Most boys and girls still have a lot of fears and guilt to overcome before they can enjoy sex and function properly.

The biology of growing up and the anxieties induced by our culture are in conflict and produce anxiety which in turn reduces sexual pleasure.

The adolescent has to pick his or her way through this sexual minefield and it is scarcely surprising that some are blown up. More rational attitudes could either increase or reduce teenage sexual activity but they would certainly inculcate a greater degree of responsibility into adolescents. Rebellious sexual activities would be greatly reduced; anxiety would be lessened, and youngsters would have a healthier and less confusing start t their adult lives.

ROMANCE

Romance is, or should be, the mainstream of our emotional loves but not the whole of them. Romance is valuable and we should all try to sustain and nourish it if only because evidence suggests that it is very important to women. So men should not allow their romantic side to wither over the years and women should be more open in revealing their romantic needs. Only in this way can a couple keep the flame of romance burning over the years.

PARTNERSHIP/MARRIAGE

Sex partnership is the framework within which most of us express our sexuality for much of our lives and it is a relationship basically between personalities rather than genitals. Our sex organs were designed to work without constant attention and fussing. They are simply used to express our personalities in a genital way. We express our personalities in other sexual ways all the time.

Any close man-woman relationship offers a whole range of wonderful options as a way of life. Provided expectations are not set too high an the couple keep their feet on the ground a partnership can be a:

- friendship;
- source of attachment;
- alliance against a hostile world;
- source of companionship;
- mutual admiration society;
- therapy group of two;
- work group with each what they do best;
- source of tender loving care;
- means of keeping romance alive;
- secret society with its own language nd history;
- child-rearing group;

Love can then be increased and can put someone else first in life.

SEXUAL ANATOMY

Genitals tend to have a bad reputation in some cultures an frequently their existence is completely ignored or denied in child rearing. They are often called 'private parts', but as human beings come in one of only two genital forms this description seems a little misplaced. Our mouths secure our persona survival and our genitals the survival of the species.

Genital also help to improve the relationship between the sexes. Therefore, a reasonable and unobserved knowledge of and acquaintance with sexual anatomy and physiology is both healthy and helpful.

COPULATION

Knowledge of the sexual arousal mechanisms can be helpful to enable both partners to know that each is properly aroused and exited by what they are doing during foreplay and intercourse. For example, a man skilled in his partner's sexual response can tell by feeling her nipples or clitoris exactly where she is in the cycle and so know how best to caress her or whether to go on to intercourse. Obviously a preoccupation with bodily changes at such a time is unhealthy and unloving but a little knowledge, practically applied, can improve the quality of one's foreplay and intercourse enormously.

TEACHING CHILDREN ABOUT SEX

Sex education should prepare the young for the conflicting and difficult emotions they are about to experience (or actually are experiencing) and should help them cope with them. Adolescence is a difficult enough time for parents and children alike and to add to their troubles by ignoring the real problems and simply teaching the bald facts is to do them less than justice.

An ideal sex education programme would undo the previous harm, correct tendencies towards perversions and direct children who need more personal education and care or perhaps even therapy. Through insight it would simultaneously maximise both the child's potential and control – these, surely, are sound educational aims whatever subject is being taught.

MASTURBATION

It is difficult to escape the conclusion that if masturbation, in the full psychosexual sense, proceeded with less difficulty in adolescence, then intercourse would be improved to the benefit of the man-woman relationship. Opposition to masturbation may have made sense to some people in the past as a control against sexual expression becoming rampant, but there is no justification for repressing it today. In fact, the reverse is the case; those individuals who are most accepting of their sexuality in all its forms are the ones who are most responsible about it expression.

HOMOSEXUALITY

A homosexual orientation can probably result from a large number of factors. It is said to be rare in communities which do not make an endless song and dance about heterosexual behaviour. In any reasonable society an individual's sexual orientation would be his or her own concern. Today, the majority of the population probably accept homosexuality, believing that it is up to the individual to make his or her own choice.

SEX IN OLD AGE

Increasing research indicates that a good and continuing sex life is beneficial for the older citizen. As at all ages the most serious detrimental factors are inhibitions instilled in childhood. They affect the sexual behaviour of the elderly more than age itself.

Sex in old age has received little attention, or indeed encouragement. Obviously much more could be done in this direction, especially as there are now many millions of elderly people with increasing expectations in the Western world. The problems are heightened and made more poignant because of the numerical imbalance of the sexes over the age of sixty. Anything that could be done to help the growing number of elderly to enjoy their lives and fulfil themselves better in this area would probably be welcomed by most people of all ages.

LOVE

From the point of view of attachment, a mature adult is one who can both give and receive love. Life, to be successful, depends on maintaining a balance of dependence on and independence from others.

A mature person can ask for love when he or she needs it and, knowing that he or she will get it, feels confident to give love to others. It is difficult and probably impossible to extend love to others in a mature way if one has not received it or is receiving more oneself. To this giving and receiving of attachment love genitality is added in adulthood and this further deepens and strengthens the loving bonds.

There is no more 'ideal love' than there is an ideal marriage. We are all complex, ever changing beings whose ability to give and receive love varies from day to day and from year to year. What a shame it is that more couples do not realise this as they cast around the sexual arena, or go for professional help in an effort to improve their lot when in reality what they already have is pretty good.

In matters of love, do not commit the grievous error of making the best the enemy of the good. Always remember that the ideal does not exist in this world.

5. IDEALISTIC HUMAN REARING
RAISING A CHILD

In raising a child, there is no reason why prospective parents should be reminded of the simple rule of love and care for their offspring. This being a basic need, it is taken for granted that everybody is well equipped for this responsibility. Loving a child is one of the first and foremost of human motives. In most of us, it comes natural. We wish for the conception, we want the best for the child and we crave to hold the little one in our arms. Many feelings go with love, affection, warmth and positive thoughts. From the moment of conceiving, all the comforts of the pre-natal stage in the mother's womb and the early stages that follow the birth.

Often forgotten or neglected, is the father's positive contribution and feelings that accompany the mother's love for the unborn child. The child's experiences in the womb and the subsequent post-natal caring form the basis and the beginning of a great loving relationship between the child and the parents.

The pre-natal stage and the post-natal understanding of the child's needs, help enormously in the formation of the child's character. At this stage, the development of the child's personality, his/her education, the understanding of values, all this become predominant in the parental home environment. This requires the contribution and the effort of both parents and whenever, if possible, the attention of the extended family.

Such care for the unborn or newly-born child, requires a little bit of planning and preparation for the big event of giving birth and of preparing the child to enter a chosen society. It cannot be disputed that the growth and the upbringing begins in the mother's womb. Even less arguable, is the fact that the parents need maturity, knowledge, understanding, training, and guidance in undertaking such an important responsibility. Social demands on the child and the parents are such that some reminding to rearing a child becomes a necessity.

Parents should not feel guilty whenever they seek help in understanding the child and his/her upbringing. Such an action should be considered wise. Guidance towards parenting should be readily available, in clinics, in surgeries, in counselling establishments, in bookshops, in academia, on bookshelves and on the coffee table.

There is no age limit in asking questions and in enquiring as to the best method for caring for your child. The reward for loving and caring for a human being is the most satisfying feeling ever described. To feel that you have done your best for that child is a personal sensation that nothing else can compete.

To perform well as a parent it may call for some form of guidance. If it means reading on how to give your love and offer your care to your child, so be it. There is no reason for holding back your love and care and certainly no room for quilt regarding your reading on such an important topic. Parents spend a great part of their lives working and preparing for a family. Any opportunity given in parental education ought to be pursued.

CHILD CULTURE

The human race, in its evolution, has made a very slow progress. All the way it has had to make experiments in many ways, sometimes succeeding and taking a step forward and on many occasions experiencing failure and re-trying. Where people have gained by their mistakes, failures have in a way resulted in progress, but where failure has not been overcome, the same mistakes have recurred.

Political and other leaders, mainly in education and religion, have been advocating for many years that the cause underlying most of the difficulties human beings experience as adults, is the wrong training received in early childhood. While the leaders' claims appear reasonable, little improvement in the things which make life most worthwhile is apparent as the result of all kinds of efforts.

Achievements in all forms must be judged by the benefits which they bring to the human race and if the methods of child upbringing are not producing a greater measure of freedom, prosperity and happiness throughout life, then something must be wrong with the system of educating and caring for the child.

Therefore, the system of child training must be reviewed and examined at every step, to find where the failure lies. As the popular education is looked into, there is much that is commendable and perhaps a few radical changes could be made without altering the entire method of educating society. It may become obvious that the early years of the child are the vital ones.

It may appear logical that any failures cannot be attributed to the child and that advising counselling and educating may not be for the child, but rather for the parent, the guardian, the teacher, the adult. For in their hands lies the future of the child, the future of the human race, the next generation.

Psychology, some say is the science of human behaviour, or the study of the mind and out of the psyche in its various sections, grow all the feelings and emotions. All actions and conduct of the individual depend on sentiments - feelings and emotions.

Whatever the pattern of behaviour, whether from a mental, or material point of view, the attitude of the individual is determined by the action of the conscious mind and its reaction upon the subconscious.

Astonishingly, although it is well known that the mind is behind all actions, little attention is given to the study of the mind and its relation to life in general and the environment.

If a person wants to learn about agriculture, he/she is expected to know about the soil and the best method to yield the maximum possible. A car mechanic must know everything about the car and he/she is expected to undergo the appropriate apprenticeship and training and subsequently gain a lot of experience in the field. But, people in general give little, if any attention to the functions of the psyche, which they must use during the whole of the existence on earth. The skill in the use of the mind determines the success or failure in life.

Whether a farmer, a computer scientist, a teacher, or a labourer, the mind is behind any actions and in performing one's own duties. Hence, the importance of the study of behaviour and human sciences must be remembered. Just as important, being the relationship of Homo sapiens (being the major intelligent manipulators and participators) to the whole of the environment. Not forgetting the effect of the human actions to the surroundings and the influence on what the next generation will inherit.

In helping a child to grow, one must realise that whatever it is or does, desirable or not, is the result of the mental action. Since the action of the mind of the child is influenced by others, it must be realised how important the behaviour of those close is. Predominant the influences can be from the mother, father and siblings. Extending to teachers and everybody else coming into contact with the child.

Each generation has two distinct tasks to perform. One is the responsibility to advance the present civilisation to the highest possible point. The second task, a simultaneous one, is to prepare the generation that follows; equipping the young ones with the necessary skills, respect, knowledge and the understanding of life at large. The outcome of this is the continuation of civilisation onto a higher plane.

Engrossing one's interest on of the two responsibilities is not enough. There must be equilibrium between the two tasks. History presents many examples of civilisations that became so obsessed with arts, sciences, conquests... to the point of neglecting the training and care of the children. When children, thus neglected, get their chance they make a shipwreck of the fine achievements for which the previous generation gave everything.

Many neglected children have a tragedy to compose. The father may be a successful businessman and the mother outstandingly successful in politics, but the time comes when the children cannot inherit the full values of life. Inheriting the material successes of the parents is not enough.

Many times, these children enter an era of dissipation, squandering not only the accumulations of their parents, but much of what other generations had acquired. Such parents defeat their own purposes in rearing children who lack the ability to continue the work and ideals of the family. Parents may glory in their success and point with pride to their achievements, but as long as the children are not taught the sanctity of life and the importance of individual responsibility, little progress can be made.

The subject of child education and development is often mentioned by leaders and the mass media, but does not appear to be of sufficient importance by the law-making bodies. Certain authorities believe that it is very difficult to secure enough funds for schools and education. Teachers are earnest and sincere in their efforts for the children at school, but are handicapped by crowded conditions, lack of equipment and in a few cases undernourished condition of the children.

The cutting of school funds and teachers' salaries is something much more serious than its immediate effect. It means that the educational system and the human resources are morally low and de-motivated and as such the children, the coming generation, will not be a worthy successor of the national inheritance. It is well known that parents would rather suffer hunger than have the children do without. But legislatures approach the all important subject of child education without any realisation of the momentous consequences which they control.

The ancient Athenians, with all their wonderful ideas of beauty and equally outstanding skill in objectifying, they failed to impart these powers to their successors, the younger and next generation. With such certainty, one can say that this was the major contributor the end of the Hellenic glory. Every generation is jeopardised by the danger of entering so enthusiastically into its achievements as to neglect its childhood. A neglected childhood is always ready to make junk of the achievements of its earlier generations, the arts, sciences, technology, inventions, even of their dreams.

The child is a part of the human race and is attached to the parents. To attempt to understand the child as a unit by himself/herself, without taking into consideration all the influences brought to bear upon her/his life by parents, teachers, religious leaders and others, would be as great a mistake as to try to comprehend the value of a car by looking at the shape alone. The support of the child would not be complete if attention is only given to the appearance of the child. If the child is to have the finest of development, in its holistic sense, those who bring influence must be considered, especially the mother.

There is a very strong bond between the mother and her child. Her influence upon the child before birth is great and reaches back, even more, into the life and character of the mother. This primary influence of the mother, together with many other factors having a part in the early life of the child, must all be taken into consideration in bringing up a child.

Carefully regulated habits and a high standard of conduct for the child are not sufficient to prepare him/her for life. These are of value, but they are only the outer shell which contains the major characteristics of the child.

The different periods of development must (each stage of development) be given the consideration it deserves. Psychologists divide the development of human beings into parts. Simply speaking, the influence of the mother and others on the individual expression of the child prior to birth and until emotionally grown. This is known as the pre-natal period of growth. The second part of development is the post-natal stage up to toddler-hood. During this period, there is a tremendous influence of the mother upon the child, particularly if the child is breast-fed. These first years are of inexpressible importance in the development and formalisation of the character of the child.

The third part of the child's development extends into adolescence. The child's main functions are based on earlier experiences and the influences of the present surroundings. This is the time for the parents and those near who are interested in the welfare of the child to sow seed which they wish to see come fruitage later, for whatever is planted in the soil of the child's mind in these early years is sure to bear fruit.

Later in this period the child passes through a time of great importance in which physically and mentally he/she is being prepared for a change which becomes manifest in the adolescent period. If understanding and tact are used in the direction of the child through these years, the child will be well-prepared for the next stage of his/her journey toward maturity. Many questions are in the mind of the child as he passes from early childhood into a larger world where there are many things difficult to understand. Much wisdom, great love and patience are required for the safe passage of the child through these years.

During adolescence, care must be exercised to see that no harm comes to the child while she/he is bringing the gap from childhood to manhood or womanhood. These are precious years and to fulfil their purpose will require consecrated effort and interest on the part of the parents and teachers. It is sometimes called the awkward age and the child is laughed at because he/she cannot gracefully accomplish all the re-adjustments that life seems to demand at this time. During this period children are particularly sensitive and their pride is easily wounded. They may become sensitive and uncommunicative when in the presence of grown-ups. Empathy and understanding, based on one's ability to remember her/his own youth, will form a bond between adult and child at this time which the years cannot break.

The next phase of development is the ushering of the youth into maturity. As the child grows older, the parent, teachers and others responsible to a degree for her/him, must learn the ratio of their diminishing responsibility, which ends completely when the child arrives at absolute accountability for reaching his/her own conclusions.

PHYSICAL ASPECTS

It is claimed by some religious leaders that if the child is given into their care exclusively until the age of five years, they can absolutely determine the convictions of that child and, to a large extend, his character for the rest of his life. Where this claim may bear some truth, there is more vital a period, even before the first five years.

The most vital period in the life of the child covers the nine glorious months in which the mother carries the precious life within her womb. This is the mother's reign, a time when she rules absolutely in her kingdom. It is true she has but one subject, hers to make or mar. Mothers may not realise the power that is theirs during pregnancy, the veneration and awe, pregnancy being the greatest experience of life.

On the other hand, pregnancy is often accepted as an unavoidable evil, the child undesired and the mother looks upon this period as one of great sacrifice and hardship. No plans are made for the welfare of the little stranger, no high ideals held for the unborn child and no deep yearnings or mighty aspirations stir the mother's heart in reverent meditation. Nurtured in loving thought, rejoiced over in song, the child cannot but be great.

The hand that rocks the cradle rules the world, but most of the work has been done before the child arrives to occupy the cradle. The vital period is before birth. The mothers must realise their power and exercise it for the good of the child.

Pre-natal influence is now an accepted fact. In the past, hideous sights witnessed by the mother during pregnancy, or tragic experiences of the mother at that time, were blamed for birth marks and other abnormal conditions appearing in the child at birth. This is seeing only the negative side of the picture. It is true that there have been many individual cases where the mothers understood something of positive thinking and exercised it for the good of her unborn child with marvellous results.

The question may be asked, why this tremendous influence of the mother upon the unborn child? It is generally accepted that what is known as the subconscious mind carries on all the functions of the body. Therefore, the growth and development of the child within the uterus is under the direction of the subconscious mind of the expectant mother.

This must be true, for the child has physical connection with the mother and this connection continues until severed after birth.

When the child is born we have a continuation of life which was. Hence, the great importance of a proper direction of the feelings, hopes, ambitions, inspirations and ideals of the mother. To a large extent, in the beginning of the child's life, its subconscious actions will be but a reflection of the subconscious actions and reactions of the mother.

One of the great assets of life is good health, a fact which does not need to be explained or enlarged upon. Every mother desires for a child a strong, healthy body, free from defects and with strong power of resistance.

The expectant mother may not herself possess vigorous health. She may be subject to various forms of ailments and lack strength and vitality. When a woman of this type realises that she is pregnant, she should at once begin to picture in her mind the kind of child she wants. She should see in her mind a strong, healthy child, romping and playing, never ill, always happy and full of life. This mother-to-be should minimise her own aches and pains and magnify and enlarge upon any short or long periods of physical well-being which she may enjoy. She should not listen to tales of sickness or accidents and, so far as is possible, keep her mind in a happy, peaceful state.

If possible, she should associate with strong and vigorous people whose tone is uplifting, because its key-note is health and laughter. She should not be afraid of fun and jollity, for laughter is a splendid physical tonic. The pregnant woman cannot afford to jeopardise the well-being of her child by giving attention to anything but the very highest and best within her reach.

She should look at pictures of handsome men and beautiful women whose faces express strong character, for she wants her child to be endowed with good looks as well as strength and health. Beauty of face and body in an individual represents a higher expression of self from the physical standpoint. So, in her dreams of the child-to-be, let the mother build beauty, strength and health.

This reading will not be complete without a few suggestions on diet and personal hygiene. The mother will see that her menu contains plenty of fruits and vegetables, for they are the foods which contain the valuable mineral elements which are so important in building a strong, symmetrical body. She should drink plenty of water.

Exercising, such as walking, or any other light exercise should be taken daily. Housework, if it is not too heavy and exhausting, is good, but the expectant mother should spend some time outdoors each day. She should take a bath frequently, everyday if possible and take a good rub after leaving the bath. The teeth should be brushed three times a day, or after every meal. Instinctive cleanliness is a thing that is passed on to the child. Man is a water animal; the child in the womb is formed in fluid and lives in fluid for nine months. It is born in water and if the mother likes her bath and relaxes and rests in the water, the child will be much easier to train in taking its bath. Some babies tease for their baths, even before they can sit up, when they see the water being prepared; other babies not only seem to dislike water in general, but kick and fight and almost go into convulsions at bathing time.

The pregnant mother should have plenty of rest and sleep, sleeping at least eight hours out of the twenty-four. She should guard against getting over-tired and exhausted. If she is engaged in housework or about other matters, she should rest several short periods a day. Have a siesta.

PSYCHOLOGICAL FACTORS

It is a difficult matter, to discuss the physical body and the mental life separately, for man is so constituted, that he is a unit of body and mind. Every phase of his being overlaps and inter-wines the other phases. It can be said that the mind influences and controls the body and the body in return reacts upon the mind favourably or unfavourably as the case may be and may determine the next mental action. Hence, many of the statements regarding the physical well-being will apply to mental culture too.

It is important that every precaution be taken, so that the child may come into the world with a perfect physical body. If we were to stop preparing for the physical well-being of the child, we would fail far short of accomplishing our task. The mind of the unborn child is of equal, if not of greater importance than its physical body.

Since the function of the mind is to think, we must start with the mother's thought and endeavour to help her to keep her thinking running in the right channels. It is not the idea that the mother shall become a logician, delving into deep philosophical or scientific questions, but she must learn to control her thoughts and train herself to think the kind of thoughts which will produce desirable results in her child. For thoughts are creative and produce after their kind.

The expectant mother must free her mind from all thoughts of so-called evil, such as criticism, jealousy, envy, unkindness, cruelty, gossip, fear, anger, greed, misery and selfishness. It cannot be done by simply trying not to thing of theses things; it can only be done by substituting in their place the best thoughts of which she is capable - thoughts of love, kindness, peace, courage, strength, hospitality, forgiveness beauty and goodness.

The list is endless, but this is not easy. For when one is stirred by incidents which ordinarily produce irritation and anger, it is difficult to hold the mind on thoughts which represent the opposite of the way one is feeling at the time. However, it can be done and must be done if the mother is to become an instrument for good to her unborn child.

One means of accomplishing this, is for the mother to memorise some lines of poetry or prose which express beautiful thoughts, poetry and ideals, or which are simply statements of truth. These may be just one or two lines, but if they are memorised they can be used at any time.

Sometimes, feelings seem to be stronger than at other times and require all one's energy and strategy to eliminate them. If she has some beautiful thoughts memorised, she can repeat them over and over, compelling her thought to dwell upon them. The repetition of the lines can continue until finally the feeling of irritation will be forgotten for the time being and the mind will become quiet and peaceful.

It is true that the discordant feeling may return and then the work must be done again, until finally the enemy is vanquished and harmony is restored. For discordant thoughts are the worst enemy one may have and for the expectant mother to entertain such thoughts, means the destruction of any high ideals she may hold for her unborn child.

The expectant mother should read books on biography and travel and try to remember the happiest, finest, noblest incidents. When some ugly thoughts flash into her mind, as they may have a habit of doing, at once she should turn her mind on some of the pleasing pictures which she has been storing up in her mind for just such an emergency.

The power to think is so tremendous that it may well be given earnest and serious attention by any man or woman who expects to build a character worthwhile. But in its importance to the expectant mother, it outweighs every other factor in her preparation for the coming child. It is not easy to control one's thinking and produce thoughts that are of a constructive nature. It is a task worthy of the best mettle, but its rewards are commensurate with its difficulty of attainment. While the mother may not be assured that she will succeed perfectly in weeding out all undesirable thoughts and substituting only the good, nevertheless, she owes it to her child to do her best.

She must deliberately decide what she will think about. If she permits herself to think thoughts that are not constructive, she is not protecting her child, but is blasting its future with the results which her thoughts are sure to produce. Many-a-mother would willingly lay down her life to save her child from the results of its folly, when in its youth or early maturity it has become involved in serious difficulties.. These difficulties might have been avoided had the mother understood the law of pre-natal culture and practised it in her thought life, during the time she was carrying her child.

A woman's thought during pregnancy will naturally turn often upon herself, her present condition and the time when the child will be born. During this entire period she is hyper-sensitive, peculiarly susceptible to her own suggestions and the suggestions of others. So it is comparatively easy for her to impress upon herself the ideals which she covets for her child and along this line her thoughts should be directed.

If there is any fear or anxiety for her safety at the time of child birth, she should get rid of it at once. Pregnancy is natural; there is nothing to be feared, for women have been giving birth to children through the ages. Woman has been prepared for this service and her body formed as it is for this particular purpose. At this time, if she will suggest to herself strength, courage, and high purpose, it will have a splendid effect upon herself and also the child. The percentage of cases where there is serious trouble at child birth is much lower than we have been let to believe and in many such cases, this trouble could have been lessened greatly if not avoided altogether, by the practice on the part of the mother of the principles set forth in this text.

The mother should be careful to entertain no thoughts or feelings of self-consciousness or timidity. There may be a tendency toward embarrassment because of her physical appearance, particularly through the last few months of pregnancy, but there should be no sense of shame or embarrassments, no more than there should be in the fruit tree because of the enlargement of the bud for the coming forth of the flower and fruit. If there is any such tendency in her feelings, she should overcome it and force herself to socialise, meet and talk with people.

She should not permit her appearance to keep her within doors and away from people. She should go out and mix with others. But, be sure that her associates are persons who have high ideals and the right attitude toward life.

The mother-to-be should attend lectures, concerts, social events, church services and any entertainment of high order. If a woman remains in solitude during pregnancy, her child is likely to be timid, fearful, backward and retiring. Therefore, for her child's sake, the expectant mother must continue her social life and keep up her interest in the outside world.

No great attempt should be made to wear clothing that will conceal the fact of pregnancy, but rather the mother should be proud of the fact that she appears to be in preparation for the accomplishment of one of the highest and noblest expressions of nature.

She should not allow anyone to express sympathy for her because she is pregnant, or feel any regrets in the matter herself. Rather she should glory in this wonderful experience and make it count to the very utmost in her own life and in the life of coming child.

This is not the time for the expectant mother to become mentally lazy and inactive. Additional to her physical exercises, the mother should exercise the mind, as well. She should keep her mind busy along constructiveness of thinking. She should read good books and think seriously for at least a short period each day on some statements. Poems, or ideas expressed in her reading. If she keeps her mind occupied with the highest thoughts, trying to understand things and philosophical aspects, her power to think will increase and she will advance step by step in her mental development. This will increase the mental capacity of the coming child.

Regardless of appearances, there is nothing to worry about. She should think the brightest, happiest, most kindly loving thoughts she can muster into her mental realm.

By now it has been shown that while the mother's thoughts have a direct bearing on the child's life, it also has a very important influence upon her own life, both present and future.

The usual feeling of the average parents toward the coming child is one of pleasurable anticipation of having a child in the home to love and care for. They may have high hopes and ambitions for that child, but rarely do parents expect their child to go far beyond their own attainments. By their own thinking regarding the child and their low ideals for him/her, before birth, an influence which is likely to limit and narrow the child's accomplishments.

Sometimes, a woman becomes pregnant against her wishes and there may be a tendency to rebel and there are cases where a mother entertains a feeling of resentment. For her own sake, as well as that of the child, a woman in such a situation will do well to consider carefully what she is doing and bring her thoughts into line. It must be realised that the child is not responsible for his coming, that he/she is an innocent participant. The mother, by her thoughts can make this child a blessing to herself. On the other hand, by the terrific influence which she is able to bring on the growing embryo by her strong feelings of hatred, she may brand the child and doom it to a life of crime.

So, if conception is not desired, every feeling of displeasure or regret should be put aside and feelings of the opposite nature should start. It is the woman's task to find these happier pictures and build into them a mother's love and joyous expectancy and welcome.

The expectant mother should see her child perfect, happy and strong. What she expects comes to pass, through positive expectation and thought. This attitude of right thinking is all important in giving the mind of the child a good and happy impression. Avoid as far as possible, arguments, emotional scenes and anything that will cause her distress.

The expectant mother should avoid and reject evil thoughts as she would poison. She must be careful of her feelings which are the result of her predominant state of mind. As she cultivates thoughts of beauty, love and peace, these attributes will gradually become her predominant feelings and will produce, in her life and the life of her child, pleasing and happy results.

A happy relationship should be established between mother and child during the nine months of great intimacy. In future years, the mother will appreciate the companionship and confidence of her offspring.

The time to establish this close relationship, which will last throughout life, is while the mother is carrying her child. She should in mind talk to her child of her own highest ideals and any secret ambitions she may have had which were not fulfilled. She must not try to impress these ambitions upon her child for their fulfilment, but talk it over with the child as she would with a most loving and understanding friend. Then she should talk to the child of the great possibilities before it of the wondrous beauty and grandeur of life.

The mother should see her child coming to her with all of its little trials and difficulties, knowing that the mother will always be ready with her sympathy and help. In this manner a mother may prepare the way for a very close and happy companionship which will be a lifelong asset to both.

In thinking of her child in such a manner, the mother will see it playing with other children and associating with those of its own age as it grow in years. The mother's place cannot be filled by another, neither can she try to fill the whole of the child's life. She is only one of the many friends her child will require for a well-rounded and complete life. Above all, she should not bind the child to herself to the exclusion of others.

The mother should not be jealous of the love of her child. However, much the child may come to love others, if she is of the right mental attitude, there can be no question of the child's love for her.

The mother should suggest to her unborn child, an eagerness for knowledge. Should see it excel in education and with many interest and talents in life and with many accomplishments.

PREGNANCY PERIOD

If there is a picture in the home that is inspiring, the mother should spent time before it each day and visualise the ideals of which she is capable. Copies of the great masterpieces cannot be expensive and will serve this purpose admirably. Visit the art galleries and let the eyes and the soul feed on the marvel of art.

If there are flowers in the garden or home, she should spend as much time as she can with them, for they are beautiful expressions of nature.

Also, frequent visits to the parks, town gardens, to the country and anywhere where nature speaks of peace and harmony.

The home environment should become as attractive as possible. Even if the dwelling is humble, it can be made neat and attractive, comfortable to live in, indeed a home.

The mother's personal appearance should be immaculate, wearing pretty clothes with bright, harmonising colours. This is no time for her to neglect herself. Now is the time for her to establish habits of which she will approve in the future.

In her visits to public places, the mother should listen to lectures on subjects of interest. She should go to theatres and cinemas. Many subjects are elevating in thoughts. Try and read some good literature, listen to soothing music and watch some good television programmes.

A prospective mother must make her words count for good. She cannot afford to become irritated and speak angrily to anyone. Kind, loving words should be used and should acknowledge any service rendered in appreciative words. In her conversation, she should talk of things which are of value.

CHILDREN'S IDEALS

By now, so much has been explained about the mother's unlimited influence on the unborn child and the tremendous power which she wields during this period. A prospective mother can make of her child almost anything she desires. The mother may begin to plan the future of her child in detail, his profession, social standing, education, ideals, perhaps including some of her own ambitions, or some ideals which she may long have cherished.

Having set forth the mother's power along this line, it must be stressed that she does not own the child. The child is an individual in her/his own right and therefore not of possession or ownership to do with ass she pleases. It is hers to bring into the world under the best possible influences, hers to love, nurture and direct through childhood and youth, but here her rights in this matter end.

The child is an individual and no one has the right to choose for him/her the path which he/she will follow through life. That must be left for the child to do for himself. The mother has no right to impose her feelings and desires on the child. To do so, would probably make a misfit of the child and a failure of his/her life. Whereas, if she uses her influence to guide and encourage the child along the line of his own inclinations, he will most likely attain the goal which will represent his own wishes.

It must be remembered that the child being an individual, comes into the world with certain tendencies which will enable him to be more successful in some lines of endeavour than in others. Therefore, the education and training which the child receives, both before and after birth should be of the purpose of encouraging and developing these natural tendencies, for it will be along these lines that his greatest success and happiness will lie.

Hence, it is not wise, nor right for any mother to be specific as to the calling or profession or life work of the child, but the right mental attitude of the mother is to know that the child is good and successful in whatever he selects to do.

A mother with the best intentions may ruin the life of her child by choosing the vocation she/he is to follow. There are second rate doctors, lawyers and priests, who would have achieved marked success in other fields. The mother's ideals for the child should be the very highest, without being specific regarding the individualised expression of his life. She should want her child to have a fine physical body, beautiful in appearance, splendid health and individual ideals.

She should want him to be mentally alert, bright and eager to learn. But in her ideals for his education she will leave the details to be filled in by the child later on. In the matter of education, the world advances, technology becomes more complicated and educational systems change, so she should visualise the child equipped for life with the best education possible.

In the matter of the child's personality, she will want him/her to be with high ideals, honest and true in every sense of the word. This will necessitate high ideals for herself and integrity. It is important that she embody in her own life during pregnancy the virtues which she wishes to see depicted in the life of her child.

Spiritual values cannot be ignored in the life of an individual, in favour of wealth, power and material achievement. In her thought of him, a mother will want him to be of great service to his fellowman, whatever line of work he may choose to follow. Success is to be desired, but it means more than the accumulation of money. A mother must include these in her daily periods of relaxation.

Practically, all study of Child Behaviour takes into consideration only the conditioning and the physical body of the human being, to the almost complete neglect of the spiritual world of the child, which is as vital and all important part of the human being for it determines what his is in life, his health, success and happiness.

A discussion of religion may be out of place in a book of this kind, but it is impossible to present any method or system of training for the child without an adequate understanding of this important truth, regarding his spiritual nature. It is not the intention of the writer to go into the matter of religion as such, for the mother may or may not be religious according to the usual conception of that term.

Nevertheless, everyone is more or less religiously inclined and the prospective mother will do well to develop her religious or spiritual nature during this period, even if it is for the child only. This does not mean that she will think of herself as a 'sinner' and count her numerous wrong doing. Avoid quilt or any fear for the future; emphatically no quilt feelings. Whatever the experience, is in the past and nothing can be done to change it. Make good use of the present, now.

In the quiet time of her day, she should think of some of the wonderful achievements of this age in science, invention, progress in technology and computing. Then try to realise that progress will continue, into the future. If the mother wishes her child to be the channel for high ideas to be used for the advancement and improvement of mankind, she will keep her thoughts as much as possible on progress.

It is not intended that the mother should be up in the clouds all the time, but if she will endeavour to see the good in everything, the everyday facts of life will assume a brighter aspect and she will be able to move through a day of distracting duties, calm and peaceful and with confidence in the future of her child and all things in his surroundings.

Again, it is urged that she avoids anxiety, worry and all such destructive thoughts. If she persists in positive thinking, she will be able to get rid of the undesirable thoughts, simply because they are receiving no attention and they finally disappear. As the mother relaxes and keeps her thoughts on love and other positive things, she will have fewer thoughts of fear or worry and will gain confidence, strength and courage for the overcoming of all the difficulties in her life.

HEREDITY AND PATERNAL INFLUENCE

So often, there are undesirable traits of character in the near relatives of the parents, or there may be diseases of one kind or another back in the families of either or both the parents. The mother may have some fear that her child may inherit tendencies that will handicap the child in life. But she should have no such fear. She must remember that at all times she stands between her child and inherited influences, that if her mind is pure, healthy and her thoughts directed to the right channels, that no undesirable inherited influences can have any effect upon her child.

Heredity is looked upon as a most powerful influence in life. Parents cannot and must not accept the inevitable, giving up without a struggle, consigning their child to failure and disease. Medicine, with the help of modern technology has progressed enormously into the realms of genetic engineering. Consultants of all related professions are as eager to help, that parents must keep confident in whatever the modern scientist may suggest. Their thoughts are always optimistic.

In other words, with the help of the modern specialist and the mother's continuous positive thinking, heredity may have whatever influence the mother and the father permit it to have. If you have any reason for worrying on hereditary influences on the expected child, seek advice from the medical specialists, as early as possible during the pregnancy and adopt all the psychological factors explained in this book. The mother's mind with the help of the prospective father can move mountains.

In the matter of pre-natal influence, the mother will use her mind and the laws of nature to produce in her child something more desirable than might be produced if no thought is given and no knowledge brought to bear upon the case.

So far it may appear that the mother only is responsible for the child. This is not so. The father as well as the mother, plays a tremendously important role in pre-natal influences. For whether or not the mother is aware of it, the father has had much to do in the creative work of bringing this child into the world and the mother at least unconsciously realises that the father sustains an important relationship to the results of parenthood.

Perhaps no one will have as much influence on mother's feelings, inspiration and mental state during pregnancy as the father. His position then comes to be a very important one. What then, should be his attitude toward the mother of his child?

In the first place, his attitude must be one of the greatest possible considerations for the mother. It will be his duty to be very patient with any whim or undesirable state of mind which the mother may undergo. For during this period, her whole physical being is at high tension. She will naturally be subject to expression of the best as well as the worst in her nature.

The father may take as a matter of course the expression of the best, but be alarmed and resentful at the expression of less desirable traits, such as being easily offended and disposed to find fault. Unconsciously, there may be a little feeling of sacrifice in the mind of the mother, because of bringing this child into the world for her husband. In the deep privacy of their life, she may be disposed to rebuke him for his willingness to satisfy his own nature at the sacrifice of her peace of mind and anticipated suffering.

The father will be wise to overlook such disposition and on the contrary enlarge on the blessings of the opportunity to bring into the world a life and personality that may make its influence felt for good and thus bring honour, joy and satisfaction to the parents.

The father will make opportunities for the mother to see the most beautiful and desirable things. He will take her out for walks and rides and feel himself duty bound to be in the finest possible frame of mind, seeing and pointing out the desirable and beautiful in everything.

During this period he will be the perfect lover and sweetheart of his wife; never in the days of courtship did he give as much attention as during this period. He should be thoughtful enough to send her flowers and shower her with attention.

In being unusually attentive to his wife during this period, he must be very careful not to show anxiety, or allow her to feel that it is just as it should be. If the thoughts of the father and the mother are what they should be, there can but come into the world, a life prepared for great achievement.

Thoughts are real, whether we give any expression to them by word, look, or action. Our thoughts have an influence upon those with whom we are associated. Particularly, this is true in the relationship of husband and wife. Therefore, it is quite important that the father allow no thoughts, or feelings of anxiety to occupy his mind for if he does, the mother will register these thoughts of which she may not be and perhaps is not aware, but harm is done.

Hence, the father must think as kindly and encouragingly of his wife in her absence as when she is present. He must not discuss with his most intimate friend any undesirable feature of the matter. Because the father's character, life style and consciousness, registers upon the subconscious mind of the mother and therefore on the mind of the unborn child.

The father should guard carefully the physical well-being of the mother, seeing that she does not overwork, that she has good, wholesome food and gets sufficient exercise in the open air. He wants a fine, strong, active child, so he will throw all of his influence on the side of the mother during pregnancy, in an effort to encourage her and help her to keep alive her interest in things outside her own home. He should make her feel that her form is more lovely in his eyes than ever before and try to eliminate from her mind any sense of timidity in appearing public.

The mother naturally looks to the father for protection from material want; he is the provider of the family and stands between it and the outside world. He may have difficulties of his own and many obstacles to surmount and the prospect of increased expenses in the coming of the child may not be particularly encouraging.

Nevertheless, it is the father's business to keep his mind peaceful and harmonious because of his influence on his wife and through her on the child. It is not enough that he looks calm and smiles while in her presence, for if he is worried or uneasy, she will become aware of it.

Strength and courage are attributes which a woman admires in a man and a woman likes to think that her husband exemplifies these virtues. If a man is strong and courageous and stands so in the estimation of his wife, that consciousness within her will be transmitted to her child.

It must be noted, that in modern times women are more independent and follow a career path of their own. Mothers often enough achieve higher duties in their working environment, earn more than the husbands (or partners) and contribute even more in housekeeping and the building of their home.

The contribution of the woman to the family surroundings, her educational standards and the strength of character is very often higher than that of the man. Even so, the support of the husband or partner during the pregnancy and during the rearing of the child cannot be under-estimated.

Of all that has been written, it is obvious that what the mother bears in her inmost character, is passed on to the child and what the mother is, may be determined to great extend by the father's personality.

The careful nurturing and prolonging of the love between husband and wife, with the cultivation of unselfishness and forbearance on the part of both, will have the effect of giving all the children of their union an equal chance for success and happiness.

Much of the counselling given to the mother, in the preceding chapter, may be followed by the father to the great advantage of the coming child and the comfort of the expectant mother.

DAY OF BIRTH

The mother may have dreamed of a day of opportunity when she will be able to do something worthwhile, placing her name in the annals of those who accomplished great things. The day of the birth of her child is a day of the greatest opportunity that could come to anyone. An inventor may come on a day when he concludes his final experiments and brings into manifestation an invention that will be of untold value to humanity. This sinks into insignificance compared to the opportunity of the mother on the day her child is brought into the world.

She has a chance of giving birth to a perfect baby. Sentiments are expressed in the celebration of Mother's Day, every year, but the mother's greatest day is the natal day. This is the day she brings into life her own child.

Much of the difficulty and suffering which women experience in giving birth to children is caused by wrong habits of thought lodged in the subconscious mind. Through the generations the race thought has connected with childbirth with terrific suffering and many tragic happenings, so the approach to childbirth is filled with fear and read. This fear of pain and trouble causes her body to become tense when it should be completely relaxed and at ease. This tension is one thing that produces pain in the delivery of the child.

We build into our bodies the habitual thoughts which predominate in the subconscious mind. Months of faith and good cheer, preceding the day of birth, the natal day, will lessen the difficulty for the mother, as well as insuring the future of the child. The mother must always remember that any thoughts of fear or dread are destructive and can only harm, while thoughts of confidence and love are constructive and uplifting.

So much has been said about pre-natal influence that the mother may feel that the greater part of her work has been done, so far as her influence over the child concerned. But in reality, her work has just begun. The relationship between mother and child is always a very close one and makes it possible for the mother to guide and train her child as no other person can.

Second only in importance to the nine months of gestation are the first eighteen months of the child's life. In the past it has been thought that the physical needs of the child were the only thing of importance in infancy. Since the child's mental faculties are not developed for several years after birth, it has been thought that the matter of environment and association during this early period has nothing whatever to do with the future of the child.

The mother takes her child as a trust of great responsibility and her work is to guard, guide and direct it so that the end of its existence on earth, it will have given a good account of itself. Regardless of her circumstances, the child during its very early life is her job. The importance of this work is second to nothing. The mother's task carries with it unlimited possibilities.

The inventor who serves mankind through contributing to material comfort and advancement, the explorer who finds new territories for man's home, artists who in pictures, song and story reveal man's deepest thoughts and longings, all of these render a mighty service to the race. But the mother's work is superior to all of these, for within her hands is the shaping of a human life which has within its power the redemption of the world.

A few mothers consecrated to the sacred service of motherhood, willing to give their all during the early years of their child's life, could within one generation, start the world on its climb to greater heights.

The natal day was spoken of as the 'day of pains' through which the mother must pass and sometimes the mother felt gloomy, as to make this experience very perilous. But it can be a day of joy, wondrous possibilities, fulfilment of a mother's dream. With her baby in her arms, the mother experiences peace and fulfilment. An overflowing feeling, that her 'cup of joy' is full.

The baby lying in its cot, prepared for it by loving hands, is feeling upon its small body for the first time the influences from the whole wide universe. It is beginning to use its five senses and in due time, through many trials, will learn with the assistance of the mother to walk upright, speak and use its other functions.

The little body and mind are very sensitive to all influences for as yet all its parts and organs are tender. The mother supplies its first food and causes its first reaction to pleasure. During the first six weeks of the baby's life, its primary urge is hunger. In the natural working out events, this desire is satisfied by drawing the life-giving milk from its mother's breast.

It is thought sometimes that a child must be three or four years of age to be taught anything of importance. Actually, the education of the child begins before birth and every minute after birth there is a struggle in the little body for growth and progress. Instinctively, it tries to use one set of muscles after the other. In reality, it begins to use these muscles even before birth, but after birth these motions begin to have purpose and desire shown.

The baby likes to be undressed and will laugh and coo with pleasure for it feels more freedom and can move its body with greater ease. It is scarcely ever really still, except when sleeping. Soon it learns to hold up its head a little as the back muscles are strengthened.

The baby should be fed at regular intervals and as quickly as possible regular habits established. The child has but one way of expressing its desires and that is by crying. Whether these desires should be complied with or not as part of post-natal training, he should be taught early in life that he cannot have his wants gratified by crying. We do not get what we want by crying for it, but we get what we want because we earned it and have attracted it to us. The sooner the baby learns a part of this important life's lesson, the better it will be for him.

The child should be bathed frequently. Great care should be taken to keep the rectum and sexual parts clean and free from irritation, with just as little handling as possible. Sometimes tight clothing causes the child to touch and handle himself.

At this stage, the mother should eat plenty of good, wholesome food which will bring milk in abundance for the baby. When nursing the child, the mother should be very attentive and communicative with the child. Whilst breast feeding the mother under no circumstances should feed the baby while in an undesirable emotional state, such as anger, hate, fear or worry. It is a well-known fact that if the mother is in a state of anger while nursing her child, her milk is likely to be unfit and the child may become ill.

The mother should keep herself healthy, exercising in the open air and taking some rest everyday. She should not allow herself to become nervous. As she values the well-being of her child, she will maintain her poise and calmness. Try at all times to feel peaceful and contented.

As the child grows in years, he must have outside influences and interests and play with children of his own age. One of the first demands of the child is for food. It is hungry and must be satisfied. There is but one natural food for the infant and that is the mother's milk. Nothing else can take its place, if the mother has properly guarded her own health and physical condition during the pregnancy.

Happy is that mother who is willing and glad to assume the responsibility and nurse her child at her breast. It is difficult to understand how such a natural and eminently satisfactory method of rearing children came into such disrepute and how mothers could be willing to substitute other prepared foods.

There are instances where the mother for one reason or another is unable to supply the quality and quantity of milk which the child requires and so is forced to rely on other means of feeding her baby. Such mothers must do the best they can under the circumstances. The normal mother is well able to take care of her child in the matter of nursing it and nothing should induce her to do otherwise.

It may be that this personal care of the child will interfere in some respects with her social life, or with her business or professional career. If so, she should weigh the child and the considerations in the balance and determine which is of greater importance. No position in society and no standing or achievement in business or professional world is worth, even a small part of the value of one child. Therefore, the mother should gladly relinquish every ambition, for the first part of the child's rearing.

Men may play the financial game and make their millions and it can be wiped out in one stock market crash. They may win fame and fortunes, but other greater achievements soon follow theirs and what they have done is forgotten. They may move the multitude with their oratory, but the world moves on and the idea which they stood may be lost sight of. These are toiling more or less with transient things, necessary, but only steps in the world's advancement.

The mother's task is the directing of a soul, which for a few short years has been given into her care. No human mind is great enough to foretell the possibilities of that soul if the mother does her work well. Can she relinquish one iota of her opportunity to weld all of her influence into a mighty power for the good of her child?

The mother loves her child, with love like no other human emotion. She will deny herself food, clothing and pleasure that the child may not want; she will care for it through weeks of illness without a thought of herself; she will even sacrifice her life to protect her child. In her feelings for her offspring she combines all the elements of the love of an animal for its young, the higher and more understanding love of the human being.

As she holds her child in her arms and feeds it at her breast, she is giving it her love and part of herself. This act binds the mother and child still closer together and may be used as an instrument of great power by the mother.

During the months when the mother is nursing the child, she must realise that what she is in her life and personality, she is imparting to the child. Not only what she is thinking and feeling at the nursing periods, but all of the time during those months.

Psychologists tell us that all mental states are followed by bodily changes and that all consciousness leads to action. This is true of desires, or emotions, of pleasures and pains and even such seemingly non-impulsive states as sensations and ideas. It is true of the entire range of our mental life. The bodily effects in question are of course not limited to the voluntary muscles, but consist in large part of less patent changes in the action of heart, lungs, stomach and other viscera, in the calibre of blood vessels and the secretion of glands.

Since the above is true, the mother should be most careful of her mental health. She should avoid anything which is negative or discordant. Many children come into the world, poor and continue in that direction, because the mother has a sense of lack and poverty. She has permitted her mind to surrender to the appearance of poverty which may surround her at that time. One may have nothing in worldly goods and yet not to be poor; she/he may be happy, contented and fell that he/she has everything.

All negative and destructive thoughts of the mother bear fruit in the life of the child. Later in life, if undesirable conditions manifest, the mother may not realise that she may in part be responsible.

In many respects post-natal culture is as important and necessary as pre-natal culture. If a child has not had the proper pre-natal influence, much can be done for him after birth through ideal post-natal suggestion.

The mind of the child is very impressionable. As the years go by, these impressions become fixations. Therefore, a child may be moulded for good or bad, for greatness or mediocrity. How important then, that parents, teachers, preachers and doctors understand this and register only good impressions. Principles, ideals and other characteristics may or may not become active in the early years, but they are never lost.

CARE FOR TH E NEWBORN CHILD

The arrival of a baby is an event in any home. It is an unusual family that does not become engrossed. This is where a word of warning is necessary. The fact that the baby is sweet and dear makes it difficult for the parents and others to keep from fondling it all the time. Some babies are treated as though they are special playthings for the rest of the family.

The baby should never be on display. It may be seen by callers for a moment in its own room but not disturbed or handled in any way by visitors. During the first weeks and months of the child's life, the greatest need is for peace, quiet and love. These the mother will supply with loving assistance of the father.

The parents should not rush and hurry about the child, shake it or jerk it roughly. There are times of course when the parents will want to kiss and cuddle the child. If for any reason the mother does not nurse the child, she should by all means feed it herself.

This does not mean that the mother must give up outside interests and remain by the child's side all the time. If the child is well trained from birth, it will be content and quiet by itself whether awake or asleep and with someone to watch it occasionally to see that all is well. The mother may get away for short periods and thus keep her contact with the outside world. She should keep any avocation which she enjoys and which enables her to express herself.

It may seem that the almost constant care of the child, which is outlined as her task, would prevent her doing anything but watching it. The mother should never permit the child to demand her presence all the time. She should train her baby to be alone certain parts of the day and by resolutely remaining out of the child's room establish that habit, of the child learning to be alone, occasionally.

This will allow the mother to carry on her own affairs and not to neglect the rest of the family on account of the newcomer.

By now, the mother deals with the child in an intelligent, conscientious manner with certain well-defined objectives in mind. In working out these objectives, the feeding time brings one of the greatest opportunities. In the first place, there must be regular periods for feeding the child for two reasons; First, regular feeding will ad much to the development of proper digestive action and second, the child will begin to form habits of regular, orderly action in life.

Before feeding the child, the mother should sit down quietly and collect her thoughts. The mother must bring herself to this time with nothing else in mind except the welfare of her child.

She will hold the child carefully, gently and lovingly in her arms. The holding and the thought of this embrace will have a tendency to express her love. During this period she will give all her attention to the child. She will talk to the child while it is feeding.

These points may seem strange and unusual to the person who is not a trained child psychologist, but those familiar with psychological principles will know that such loving actions become effective in the life and personality of the child.

If for any reason the mother has to bottle feed the child, the same principle are followed, as those of breast feeding; holding the child in her arm and keeping the bottle next to her chest, with full attention to the baby and together in solitude.

Some may think that keeping the child by itself, or being alone with the mother during feeding, will tend to make it timid and unsociable in its nature later on. No such feeling need be entertained for the reason that this will only be practised in the very early infancy of the child. As the baby gets more accustomed to his/her new environment and adjusts itself to new surroundings, then the child may play with other children and in every respect lead the life of a normal healthy child.

At this stage of world's advancement, it is not possible to train entire families, or communities into the correct methods of caring for the new born child in a way that will ensure its successful progress through life. Any training must seek to develop the individual child.

There are cases where the parents had insufficient knowledge of parenting and the lives of boys and girls were ruined and their chances to become successful destroyed. Parents must learn to let the child go - become free, gradually inter-mix with groups of other children. Nothing is more pitiful than to see a grown man or woman dominated by a mother who refuses to realise that the son or daughter is a child no longer and must assume his or her place of responsibility in the world.

If the mother is to be able to grant the child freedom when he is grown, she must begin at the cradle. It must be realised as early as possible that the infant is not a possession, rather an individual human being with a purpose in life, which may be different and far removed from anything of which the parents can possible conceive. If the child is to be prepared for the next generation, this means that the child needs a more advanced and broader outlook in life.

The love for the child will doubtless be the consuming passion of the parents' life. Their love must be used to bring him/her freedom and open wide the door for him to pass on to great achievement. The child may be destined to deal with affairs and conditions which are beyond the conception of anyone living today.

The world today needs leaders of worth and courage. Imagine what the child may be doing in, say 25 years. It is up to the parents to start thinking of the education and training of the child, now. It is for the parents to prepare for their sons and daughters to be the leaders of the world's progress tomorrow. This is the parents' task. Not a paltry thing of rearing a child to be a citizen of the world, who will bring all of his influence to bear in the cause of right and purpose.

All of the ills and disturbances from which humanity suffers are the result of low ideals and lack of understanding on the part of the individuals, for humanity is made up of individuals.

If the parents of this generation can redeem humanity from the present degraded and materialistic conception of life, then it will be possible to develop mankind to bring about peace on earth and good will among men.

POWER OF WORDS

The conversation with the child should not be in baby language. The family of the child must remember that the subconscious mind of the child is accepting what is said and the manner in which it is said can be a strong impression on the child.

Parents may think that the kind of language used in the presence of a baby does not matter because the baby does not understand. It is true, its conscious mind does not comprehend, but the feelings which produce any type of conversation are registered on the child's subconscious mind. The parents should know that they must be careful in speech and conduct in the presence of the baby, as they would be in the presence of an older child.

From the time of birth, the parents should choose the very best and purest language. They should speak the words clearly and enlarge the vocabulary. If the parents speak kindly words to the child, they will find a little later, that the child will be addressing his playmates in a similar expression.

The child should be given a dignified name and called by that name. It should not be called 'baby', or nicknamed. It should be called by its proper name. From the very beginning the child should be made to realise that it is an individual human being and of importance in the world; that is it has a name the same as grown ups and is entitled to be treated with consideration and understanding.

It is never too early to train the child. Training, however, should be almost entirely by example and by word. Where children are unruly and bad - disobedient, it is usually because the parents are weak and negative in their inter-relationship with the child.

When the child needs to talk to somebody, he will turn to the parent who has always shown good judgement and steadfastness. He will feel that the parent can help him through his time of difficulty.

For the parents to be firm and positive does not mean that they will use their strength to break the will of the child. There will be little difficulty along these lines, if the parents from the very beginning are constructive in their training, wise in their guidance and good in the use of their vocabulary.

The parents should very clearly make a companion of the child, talking to it intelligently, as though it were an older person and could understand everything they are saying. They should speak perfect English (as much as they can), speaking slowly, distinctly and clearly.

The parents should talk to the child about many things, including discussions about their own family, other children, places, the world, their professions, the arts, technology, religion and philosophy. Above all the parents must remember to answer the child's questions truthfully.

Parents must agree with each other about subjects of discussion in the presence of the child. The time to settle a difference of opinion is when the parents are absolutely alone. If parents take a given position on some question and later on the child discovers that they have changed their opinion, it may then be told that after further careful consideration, they decided on a change of program due to further factual evidence.

New and different problems in the relationship between the parents and the child will be coming up constantly as the baby days are ended and the child develops and begins to form outside contacts.

There may be fear on the part of the parents that some harm may come to the child from his association with other children, that the ideals they have tried to establish in its mind may be destroyed by children who have not had the advantage of wise training. But if the parents allow themselves to have any anxiety for the child, they are doing it a great harm, for the child will sense such feelings, although he will not understand the cause, but that feeling of fear of the part of parents may weaken and undermine what they have been working to build.

When the parents send the child out to play with other children, or to go to school, they should be doing so with gladness. They should remember that the child must gain a series of experiences in order to develop normally and naturally. It will not be rational to stand between the child and any experiences which come in the natural and normal way and which may result in a warmer character, than otherwise have been developed.

There are good and undesirable experiences and the parents must bear in mind that this is the way life goes on. The parents would, very early teach the child that whatever comes to him in life, must be experienced and learn out of it. The child will soon establish a feeling of responsibility and a sense of security with reference to the experiences in life and whatever learned for his own future. The child will learn by his mistakes and errors and parents can be there to listen and guide.

If the child makes a mistake, serious enough to call for parental punishment, this many times fails to correct the child, because it has no relationship to the error. Any punishment to be effective, must be addressing the cause for the misconduct, made to fit the misdeed. Unjust punishment should be avoided

It is the custom in some families for the mother to make a list of offences committed by the children during the day and when the father comes home from work, to make her report to him and request that he punish the children. This is an injustice to both, the father and the children. Had the mother been firm, it would have not been necessary for her to relegate to someone else the task of inducing obedience. Every act of disobedience and tendency to violate should be dealt with as it comes and settled at that time.

ANSWERING QUESTIONS

Very early, the child begins to ask questions and naturally he runs to the mother for an explanation. The mother many times has the answers ready, perhaps before the questions are even asked. Even so, many times she will be surprised by some unexpected queries. The main thing is to answer intelligently and honestly.

One of the very important the child will ask is the subject of sex. Some children are interested in sex much sooner than others. It is common for children as early as in the fourth year to ask such questions and sometimes even younger. The parents may be speaking of some place where they were before the baby came and the child may want to know if he were there as well. When told that he was not there, he immediately wants to know where he was.

A new baby may come into the family and the very sensible question is asked by the four-year-old as to where the baby came from. The parents have such a wonderful opportunity to impact one of the truths of life in a way that would satisfy the inquiring mind and protect him of wrong information from boys and girls in a vulgar way.

The parents must explain that every living thing has a father and a mother. This knowledge can be unfolded in a simple language, the wonders of the animal world and how the birds mate, build their nest and hatch their young. The truth will charm the child more than a fairy tale.

Many times it is more expedient to tell the child that a baby is coming and give him an opportunity to show interest in this great family event. Give him some information on the coming of the baby. There will be no awkward questions after the baby arrives and he will be on the solid footing of confidence and understanding with his parents.

When the child approaches a parent with questions about life, the parents can explain in simple language which he can understand, that every life, whether plant or animal, has the power of reproduction. The child will be impressed with the sacredness of life, the purity of sex and the wonder of parenthood. The parents can explain that the loving union between the mother and father started his life as a tiny seed in her abdomen and that the mother tenderly nurtured him/her there for many months with her own blood, until he was strong enough to live outside her tummy. When a child has accurate information, he will not be interested in the unclean explanation of his companions.

Sometimes, it is asked at what age sex information should be given to children. This depends entirely on the child. It should always begin when the child asks questions regarding sex. Each question asked should be answered truthfully and when the child is satisfied with the answer, let it go at that. If he finds that he gets a truthful answer and he is not told to run out and play, or told not to ask stupid questions, the next time he is puzzled about something, he will run to the parents, who he thinks they know a lot and always tell the truth.

Naturally, not a lot of detail will be given at one time, for the child takes but one step at a time and as he requires further information, he will ask for it, if there is a bond and confidence between him and his parents.

The necessity of beginning early with the child is emphasised, just as soon as there are questions in his mind about sex. If the parents begin then, they will find only innocence and trust in the child. Frankly, the information can be given step by step, as and when required.

The difficulty in the past has been to defer the teaching of sex education until puberty or early maturity. When it is postponed until that late date, its benefit is negligible, if nor entirely lost. Lectures are delivered in Colleges on the dangers of venereal diseases and Aids, the terrible catastrophes which follow and the possible violation of the law of sexual purity. But while these may instil fear into the heart of the young people, they do not prevent wrong doing.

The purpose of sex education is essentially protective. To secure the maximum comprehension of the subject, the foundation teaching ought to start before the young person seeks expression of his/her sex characteristics.

Children are carefully instructed in every other subject and phase of life, except sex. Sex is the most important of all questions to be asked by the child, for its influence in our society affects every other phase, with the power to make a success or a shipwreck of life.

Sometimes it is suggested that the school is the place for sex education. But, a little careful thought will convince any one that the school can support what the family have already explained. In many cases the school cannot handle this subject. In the first place, many children already know much about sex and many have the wrong ideas before entering school. Others may be entirely innocent of any knowledge along this line.

This would preclude giving class or group information and to attempt to give private instructions in the matter of sex to a large number of children, would make it a tremendous task for the teacher. The time a teacher can devote to each child would be negligible, as to accomplish little. Again, it would be a difficult matter for a teacher to approach the matter in a frank way with the children coming from homes of such diversified concepts of the whole problem.

Additionally, the fact that many teachers may have had no adequate education on the subject of sex, it means that they are not qualified to deal with the child on this most important matter.

The medical General Practitioner is, also, handicapped in any instructions he might wish to give to the children of his community. He would have the same obstacles to overcome as the teacher and would have to meet the prejudices and false standards of modesty on the part of many parents.

The school is for the purpose of carrying on the education of the child along lines not easily as satisfactorily handled in the home. But in the matter of sex education, the parents are the natural teachers of the child and this work for them is much easier than for others. It is, therefore, the responsibility of the parents to lay the foundation of moral integrity in the life of their child.

One thing the school can do and which is already doing to some extend, is to give instructions in general hygiene and teach the child the fundamental truths of biology which will enable it to understand easily, the origin of human life.

There will always be many children in school whose parents have given no instruction whatever leading up to an understanding of sex life as manifested in plants and animals. Therefore, such instructions in the school are most essential. By studying plant life, the child may be lead in thought and reasoning up to the point where he will understand something of the law which brings human life into expression.

Another thing schools do to great advantage, is the setting aside of time when parents and/or children can come and listen to discussions on the subject of sex instruction for children from some one who really knows how it should be taught to little children. This person may be a teacher who specialises, or it might be a family doctor who is well-qualified to explain these facts in a simple way which all parents can understand. In fact, the speaker might be a mother who is particularly well-informed and in a position to help other mothers.

It is usually the function of the school to supply the resources. The parent-teacher association might take up some work along this line.

Parents must realise that much of the disease and unhappiness of their children after reaching maturity, results from a lack of instructions in the early years, including the answering of questions on sex.

When a child is in his teens, it is too late to give him the most effective advice regarding the evils which assail him on every side. The time to start is as a baby in his mother's arms, when every breath she may enfold him in an understanding appreciation of the beauty of the creative impulse.

The question may arise with parents, as to how they can train the child in the right way in the face of so many opposing influences, found in the close environment and while away from home. If the parents do their work well in the first five years of the child's life, they will have established themselves firmly in his confidence and be in a position to protect him from most of the harm to be met in his school and social environment.

The child cannot be protected by keeping him at home, away from other children, but there can be such a positive, strong, healthy influence at home, coupled with complete comradeship and confidence between him and his parents, as to practically ensure his safe passage over the perilous journey from infancy to maturity.

Naturally, not all questions by a child can be considered in the space of a book. The most important one of all, the question on sex has been given consideration. Another question which troubles children more than the parents sometimes is the subject God, religion, heaven, etc.

The sensitive mind of a little child is easily impressed with thoughts of fear, doubt and misgivings. In this case, great care should be taken, so that the child's conception of God will be one of fatherhood, motherhood and love. Time should be taken to explain any questions regarding religious things.

Death is another subject which puzzles the child and it is important that very early he has a constructive understanding of it. Otherwise, he may carry throughout life a phobia of death. The answer on death a parent gives depends largely on the individual philosophy, whether a person goes to paradise, or life as we know it ends here on earth, relies totally on the belief of the parent.

If bereavement occurs in the family, of someone very close to the child, it may be possible to handle such a trauma with the help of a counsellor specialising in bereavement cases. It is always wise to seek the help of an individual who is professionally accredited and where empathy can be shown.

In answering the questions of the child, the parents should be patient, sympathetic and understanding. Like any other question, this question is of very great importance to the child who is trying in his childish way to think through some subject which has come to his attention.

The parent who has time to stop anything she/he may be doing and lead the little mind on in its search for information and knowledge, will later on, find her child when grown to youth or maturity, still coming to her/him with his difficulties and problems. It is the parent's interest through the early years which builds the bridge that carries them together into the future of confidence and companionship.

OVERCOMING DIFFICULTIES

Children sometimes are difficult to control and they do not wish to obey their parents. In an effort to protect the child from certain dangers, parents resort to an appeal to fear, to secure compliance with their directions.

Many children are afraid to go around corners, for the bogey man is there, he is told. He must not go off the porch or a policeman will get him; in the dark cellar or closet lurks the monster and so on until the child is beset with a series of fears of every conceivable nature and to these fears can be traced many serious difficulties and failures of later life.

It is quite customary for the parents to read fairy stories to the children. Great caution must be exercised here for, while there are a few fairy stories that may be read with safety to a child, many of them are not desirable. In a fairy story very often, while there is a beautiful character, doing splendid deeds of courage and honour, coming to the rescue of someone in distress, there are other characters portrayed whose actions are evil. In the stories, these evil ones some times carry away little children and hold them captive amidst the most alarming and frightful surroundings. Such a story builds up phobias in the child.

While the child listens to these tales, fascinated, he is filled with fear and terror. When the child goes out in the dark on his own fear and panic fills his heart. Venturing into the garden, he expects to see a hideous monster approaching him, resembling the one he has been listening to.

This fear and horror of the dark, of ghosts and haunted houses become so developed in the child, as to remain even after he grows into manhood.

The child must be taught from infancy that there is nothing in the universe to fear, nothing in the whole wide world which one should blindly be afraid of. By reading bad stories, the parents instil senseless fear into the children. Unwillingly, the parents are transferring the mental life of their children from almost the 21st century, back into the dark ages, where superstition and stupid phobias reigned supreme.

If a child seems to be afraid of certain things, as quite frequently is the case, steps should be taken at once to remove that fear. This cannot be done at once, as it requires time. The child should never be laughed at, nor ridiculed, nor should his fear be made of any great importance. If the child is afraid of the dark room, the parent may accompany the child in the room. On opening the door of the dark room, the parent may explain that the room is dark because the light is not switched on. Turn on the light and the darkness will disappear.

In most cases the face of the child will light up and smile. As and when the opportunity arises discuss the subject of fear with the child and explain in logical sentences, in a matter-of-fact manner and in a calm voice, how natural everything is and how everything can be seen rationally.

A child should never be taught to fear a policeman. He should be taught that policemen are there for the protection of children and if in difficulty at any time, he should go at once to a policeman. Officers are very kind to children; they are parents themselves. If the child is still afraid of policemen, whether this is due to the uniform or because they heard unsuitable stories, the parent/s should take the child for a walk and visits, which would include a visit to a police station or to meet a policeman on the street.

Familiarising the child with the thing he fears will often banish that fear. Often the fears of the child are but the reflections of the fears of the parent/s. In their feelings the may fear many things and the child who can be very sensitive will express fear of a most unreasonable nature.

Love, confidence, faith and hope are splendid antidotes for fear, doubts and misgivings, If, the parents build their own character with such attributes, the child will automatically identify himself with the parents and become peaceful and poised, with courage and strength.

CHILD'S HEALTH

The enlightened parent knows that practically all the aches. Pains and illnesses of children may be prevented by wise care and supervision during infancy and that the parents' own attitude will tend to bring about splendid physical health in the child. A strong, healthy child with great vitality will have the power to resist practically every disease which is common to children.

The Nation Health Service still has some of the best services in the world. The parent, by listening to the Health Visitor and by attending the various clinics will find most of the answers and queries they may have. The nurses and the family doctors are only too pleased help with any uncertainties on vaccinations and preventative medicine.

A large percentage of sick children can be saved if parents understand health and refuse to accept appearances as the real facts of life. If in doubt and if further assistance is needed, ask the clinic personnel. They have been trained and most of them have been practising for years. They know how to attend to individual problems, such as the parents may be facing.

On the other hand, it must be remember that a happy environment free from fear, anger, discord and expressing love, gentleness, courage and faith, will act like a tonic on the child and tend to restore to health the one who is ill.

It is not at all necessary that a child be sick and parents should not for one moment believe in sickness. Psychologically, they should only believe in health, think health, live health, talk health and expect health in themselves and in the children. Health can be one of the strongest virtues of their family.

Parents should not show too much concern during a child's illness, nor should the child be unduly petted and pampered and permitted to rule the household just because he is ill. While doing everything possible for him, the matter of his illness should be treated with apparent unconcern and in a matter of fact way so that it may not make too much of an impression on his mind, which might lead to a continuance or recurrence of the experience.

The habit some parents have of producing a thermometer and taking a child's temperature every time his face is flushed, has a very bad effect on the child, for it is a suggestion to him that he is probably ill and that his parents are uneasy concerning him. Nothing could be worse for a child than such attitude.

If parents could avoid all talk concerning sickness and discuss other matters of interest, the children will think little of illnesses. If it is mentioned that there is a so-called epidemic in the town or community, or some child in the neighbourhood is ill, then it is the duty of the parents to take such attitude as will be constructive and helpful to their children. They should never permit the negative side to be emphasised, but state that the child in question has a strong body and will soon recover. They may state that everything will work out for its good. Such conversations in the family circle will have a strong influence on the lives of the children and if the parents talk freely along these lines, their children will come to accept these statements as the truth.

Instead of giving the child a tablet and letting it go at that, the importance of observing a healthy diet, exercise, rest and mental calmness must be observed by the parents and explained to the child. All fears and worries about diseases should be banished. The child will sense this confidence on the part of his parents and respond to it in a wonderful way.

HABIT FORMATION AND PUNISHMENT

Regarding the subject of punishing a child, there has been a great change for the better during the last few decades. Fifty years ago, the average parent regarded the period of six to twelve years as the time to break the child's will - whatever that meant. The cruelty inflicted in the process, even hinted of the tyrants of the Inquisition. During those years, the child does not always choose to obey commands blindly and some times in rebellion, declares that he is not going to do a thing. The old fashioned father in response had his whole being surcharged with the responsibility of fatherhood. He felt the eyes of everybody around him, watching to see what he was going to do when his authority was so challenged.

Imagine the lad, seized and hustled to the shed. The closing of the door spares us the cruel sight that ensued cruelties, which were prolonged until the boy had promised to obey his parents in all matters whatsoever, whenever and wherever. These promises were sweet music to the father's ears as he came out of the shed with all the fatherly pride and self-commendation reserved for heroes on.

He was too blinded with self-congratulation to see that he had done his child a gross injustice and that far down in the heart of the child, he had engendered a hatred that would grow until it would have the power to change and even destroy some of the most beautiful things in life.

When the modern father faces a similar situation, he realises that his child instead of requiring to break its will and rush to enact the shed scene, which will be ashamed of all his life, he tasks his ingenuity for a constructive solution to the problem, which will help the child.

Any talk to the boy, or girl, must be tactfully handled. If it is given in a form of lecture, it will do more harm than good. Somehow, the child must be drawn into the conversation, making interesting enough to lead to questions or an expression of the child's point of view.

Socrates taught his students by asking and answering questions, which is the ideal way of imparting knowledge. If one has sufficient interest in a subject to ask questions, his attention is assured while the answer is being given. Children from eight or ten years on, think much more deeply about life than other people realise. Information and constructive suggestions from the parents, relative to the important facts of life, will take root and bring forth fruit.

The parents should make every effort in the early years of the child's life, to establish a strong friendship between themselves and the child. This is not difficult, if the parents are sincere in their desire to meet the child on ground which is familiar to children. The difficulty usually is that the parents expect the child to look at matters from the adult point of view. This is impossible and may build a barrier between child and parents, which cannot be surmounted unless the parents are willing to concede their viewpoint in favour of the child.

Also, the parents must be willing to give some time to the child aside from providing food, lodging, clothing and schooling for him. The basis of any real help is companionship in which there is respect and confidence in both sides.

Whatever the character of the individual on whom the child depends, that person is the measure of his influence upon the child, whether or not, the parent, guardian or teacher realises this fact. It is highly desirable to direct the child in a certain way and eliminate traumatic habits.

Much of our education is benumbing and instead of developing the thinking abilities of the child, it dulls them. There is no power in the universe like the power of thought. Its skilful use has brought humanity many social developments, be it political systems, religious beliefs, scientific methodologies and artistic expressions. Parents and teachers should teach children how to think.

By nature, the child is a thinker. This is indicated in many questions it asks. Many of these questions imply that his thinking is based on standards and the truth in all situations. In many cases, teachers and parents dictate silence to the child, especially when they cannot answer the questions raised by the child. Thus, he is denied the right to think and express his thoughts.

What a different world this would be today if, instead of stultifying the child's desire to think for himself in the past centuries, he had been encouraged to think and search for the correct answers to his questions. In that case, the percentage of persons capable of individual, independent thought would probably be about double of what we have now. Definitely more Doctors of Philosophy and numerous research scientists.

Our system of education has produced a society of standardised human beings. This can be seen in the tendency of children and youth to do all things alike, to dress the same way, to talk using same expressions. If these young people had been taught how to think for themselves, they would have individual freedom to think and act - express their individuality. As it is they are following blindly some fragment of a truth, which they have not the ability to properly analyse and apply.

The child can be taught to think for himself, if the parents will take the time and have the patience to talk with him as though he were an intelligent person. This does not mean that the parent is to permit the child to argue about every request made of him, but it is rather the attitude of the parent and teacher toward the child, substituting respectful consideration of the child's conversation, for the usual condescending assumption of the indisputable wisdom of the grownup.

As soon as the child is old enough to express understanding and has reasonable degree of intelligence functioning, it would be helpful to teach him to observe the world around him,. The sense of beauty seen around him will be a never-ending source of pleasure in later life.

The parents must praise the child for every desirable act and deed the child performs. Instead of finding fault with the child, or calling it bad names if it makes a mistake, it should be lovingly shown a better way.

In order to overcome the possibility of developing conceit or undue self-importance in the child, it should be taught that all children are potentially great and that they have within themselves the same possibilities of being fulfilled as their own child has.

Any natural talent which the child seems to have may be emphasised and its development encouraged. Always be sure that the suggestions are suitable to the natural inclination of the child, his aptitude and interests. Needless to say, the parents may not influence the child with a preference of their own for a particular career.

PERIOD OF TRANSITION

The adolescent period is that part of life that is known as the stage when one is growing into maturity, the 'teens'. During this period of life the adolescent is not a child and yet he is not grown up. The seven years of adolescence play a very important part in life. For during these years, the personality, habits, mannerisms, likes and dislikes mature and they are usually carried on through life.

Those responsible for the training of the youth would do well to guard this period as though it were a rare and precious jewel. Parents and teachers should exert every effort to make certain that the adolescent is properly developed during the momentous era of life. The importance of these seven years is often taken too lightly by the ones who are directly responsible for the future welfare of the child.

A parent who reads this book and has a teenage child, may feel that not having given the child the care and attention, as mentioned in the earlier chapters of this book, his opportunities for being of great help to his child is past. This not true, for there is still much that can be done. In fact, it is never too late to do one's best. The parent should begin from the present, knowing that his child has imperative need of him through the period of adolescence.

This is a period of great importance from the physical standpoint, for radical bodily changes are taking place. Particular physical care and direction are required in order that the youth may pass this critical stage with his body unimpaired. Both, the boy and the girl are subject to these changes and apart from nourishing food, study and companionship, they also must have information regarding their physical make-up, the reason for the change through which they are under-going and the effect of this change on their future.

If the parents have answered candidly and honestly the questions asked earlier in the child's development (early childhood), the child will already know most of the facts necessary for the early adolescent stage. If these facts are not already known to the teenager through conversation with the parents, they should be explained at the first chance given. Otherwise the young person may be unduly alarmed.

The habit of masturbation is sometimes formed at this time. Parents should not take this seriously, for according to some authorities on the subject, it is a habit quite common among boys and girls. It is nor harmful as was formerly thought.

No matter how timid one may be about sex, parents must not shirk their duty to the child. It is a most important subject and the incorrect understanding may spoil the child's future. There are cases where the parent feels that he is not sufficiently informed or does not know how to impact the details, he should find out about the subject as much as possible and then skilfully pass on the information required.

Sex is usually thought of with reference t the reproductive system of the species only, but that is only one of the phases. It is the creative force in the body and its conservation and right use will enable the man or woman to create in the fields of art, literature, invention, or in whatever direction his/her talents may lead. The most gifted people are the most strongly sexed. The human race certainly finds sex pleasurable and any normal couple will find a lot of pleasure in making love.

At this point in life, in adolescence, the mind must be developed and reason must be exercised. The adolescent demands larger fields of conquest in knowledge, because it is natural to be endowed with larger capacity for reflection. New desires enter his life; he becomes emotional, imagines things and lives a life of dreams and illusions.

With the arrival of puberty and the adolescent period comes a certain degree of haste in the development of the mind. The mind now develops much more rapidly. It demands new and broader fields of knowledge. The child is now leaving behind it the weakness of childhood and is gradually preparing to assume the mental strength of maturity.

The teenager is likely to develop a liking for light fiction, such as detective plots and love stories. As the education becomes more demanding the youth graduate to more serious studies and very often they only become interested in text books. Some topics of ephemeral interest may be useful. Parents can make it a point of discussion at the dinner table. A variety of literary interests can be talked about.

From fourteen to eighteen years of age, is a very difficult time for the parents to handle the young person diplomatically and constructively. The parent must be very careful in his dealing with the child. Always discuss the whole situation freely and remember that the child does not exclusively belong to the parents.

The parents should try to feel that even when a son or daughter leaves home for a far away country or even close by, this is part of the young person's learning. Be around when they call on you, listen to their progress or regression and be prepared to share the experiences, whatever they may be. Remember, as parents, you should trust those you brought up, the people you trained and cared for, even before they were born, from the moment of conception.

The enlightened modern parents, when the offspring has reached this stage of development, must have already talked earnestly with him. The parents have noted the gradual lessening of holding onto the child, as the child grows in years. In their discussions with the child they incorporated the importance of education, of growing up strong and independent and of things to be encountered.

Now with confidence in the young man or woman, the parents allow for the leaving of home. Perhaps embarking onto a to a university study, another place for work, or travelling until he or she is ready to take on the life of an adult. It is not easy for parents. They will feel the empty corner of the home. It will for sure be, at times, a lonely life for the young ones, a struggle until they experience whatever life has to offer. Neither the parents, nor the young people must forget to keep in touch with each other, to continue their discussions and sharing of their love for each other.

These emotions never go away from the individual. Exchange information with each other, as often as needed. A life was shared for many years, this cannot be forgotten. On the contrary, with all the love and care given and taken, now is the time to look at the beauty and confidently admire the maturity of the next generation. With such a joyful parenting the young people who left home can but contribute to the betterment of the next generation - the improvement of society at large.

RESPONSIBILITIES REVIEWED

In the contents of this book one may have noticed the gradual lessening of the directing hand of the parent as the child increased in years. Before birth the mother's control an influence are supreme and during the first eighteen months the mother shaped the character of her child. Up to five years age her power is great. As the child gains in knowledge and understanding, gradually the parental guidance recedes and the child increasingly relies on his/her own resources.

So, all the way along in the life of the child, the parents encouraged the child to think for himself and to make his own decisions as far as compatible with the good of all concerned. Little by little the parents receded from a position of great prominence in the direction of the child.

As much as possible, all through teenage-hood and thereafter maturity, the control of the child was based on friendship and companionship.

It is a hopeful sign that the parent of today takes an altogether different attitude toward his grown children than prevailed in some families a generation or so ago. The parent today gets his greatest happiness in having the children go into the world and be successful. The greatest relationship is the one between the parents and the children. It has the possibility of lasting longer and the chance of attaining greater beauty.

It may be observed that the authority and responsibility of the parents have been gradually diminishing throughout the adolescent period and when maturity reached the child, the responsibility and authority of the father and mother ceases absolutely. Just as the child has been finding it difficult to adjust itself to the gradual increase of responsibility, the parents will find difficulty in gradually withdrawing responsibility, but it must be done. For when children reach maturity, the parents' responsibilities cease completely.

In reaching maturity, the child becomes strong in mind and body, truthful to his fellow men, starts a career and a new relationship. Parents, naturally feel just a twinge of jealousy. Simply, the parents will miss the little things they used to do for the child. But in time, gradually and with wisdom they let go, for the child is now a mature person, with freedom to choose his/her life and inter-relationships. They marry, they share things with partners and with lots of new experiences they develop even more.

As idealistic parents, you have now performed your duty in establishing the proper relationship between you and your children. The loving hand of parenthood has been ever ready to lead and direct in the right way. Your children have arrived at maturity, and they have gone out on the pathway of life for themselves. You have done your duty and you believe that everything will work out all right and that your children will be happy and successful.

END

INDEX	PAGE:
ABNORMAL CHROMOSOME EFFECTS	16
ADAPTIVE SIGNIFICANCE OF SEX	27
ANARCHISM	65
ANSWERING QUESTIONS	109
ARTHROPOD - REPRODUCTIVE SYSTEM AND LIFE CYCLE	52
ASEXUAL REPRODUCTION	12
BABY AND CHILDHOOD SEXUALITY	73
BIOLOGICAL AND PHYSIOLOGICAL ASPECTS OF SEX	11
BISEXUALITY	46
CARE FOR TH E NEWBORN CHILD	104
CHEMICAL COMPOUND OF SEX HORMONES	25
CHILD CULTURE	80
CHILD'S HEALTH	116
CHILDREN'S IDEALS	93
COPULATION	76
CROSS-FERTILIZATION	48
CULTURE AND PERSONALITY	67
DAY OF BIRTH	99
DIFFERENTIATION OF THE SEXES	36
DIRECT ANIMAL DEVELOPMENT	39
DUCHENNE'S MUSCULAR DYSTROPHY (DMD)	43
EARLY ADOLESCENCE	73
ECOLOGICAL INFLUENCES	10
EROS, CUPID AND SEX	5
FERTILIZATION, SEX DETERMINATION, AND DIFFERENTIATION	24
FOLK ART AND VARIATIONS BY SEX	10
FRIGIDITY	28
GENETIC AND CONGENITAL ABNORMALITIES	64
GONADS	38
HABIT FORMATION AND PUNISHMENT	117
HAEMOPHILIA A	42
HEREDITY AND PATERNAL INFLUENCE	96
HEREDITY AND SEX LINKAGE	29
HERMAPHRODITISM	46
HISTORY OF SEX HORMONES	25
HOMOSEXUALITY	77
HORMONES IN ANIMALHOOD	31
HUMAN BEHAVIOUR	55
HUMAN DEVELOPMENT	48
HUMAN DEVELOPMENT AND SEX DIMORPHISM	35
HUMAN EMBRYOLOGY AND THE GENITAL SYSTEM	38
IDEALISTIC HUMAN REARING	79
IMPACT OF PSYCHOANALYSIS	62
IMPOTENCE	45
LEGAL REGULATION	7
LOVE	78
MARITAL CUSTOMS AND LAWS	68
MARITAL ROLES	70
MASTURBATION	77
MID-AND LATE ADOLESCENCE	74
MYTHOLOGICAL EROS	5

NERVOUS SYSTEM FACTORS	60
OESTROGEN	22
OVERCOMING DIFFICULTIES	114
PARTNERSHIP/MARRIAGE	75
PEER SOCIALIZATION	71
PERCEPTION OF SEX	26
PERIOD OF TRANSITION	120
PERSONALITY	66
PHYSICAL ASPECTS	85
POWER OF WORDS	107
PREFACE	4
PREGNANCY PERIOD	92
PROBLEMS IN DEVELOPMENT	71
PRODUCTION	21
PSEUDOHERMAPHRODITISM	16
PSYCHOLOGICAL EFFECTS OF EARLY CONDITIONING	57
PSYCHOLOGICAL FACTORS	87
PSYCHOMOTOR LEARNING IN SEX	47
PSYCHOSEXUAL DYSFUNCTION	63
RAISING A CHILD	79
REPRODUCTIVE BEHAVIOUR – DISPLAYS	51
RESPONSIBILITIES REVIEWED	123
ROMANCE	75
SEASONAL OR PERIODIC SEXUAL CYCLES	33
SEX AND DEVELOPMENTAL PSYCHOLOGY	73
SEX AND THE EFFECTS OF ENVIRONMENT	35
SEX CELLS	13
SEX CHROMOSOMES	14
SEX DETERMINATION	12
SEX DETERMINATION	19
SEX DIFFERENCES IN ANIMALS	18
SEX DIFFERENCES	13
SEX HORMONES	21
SEX IN OLD AGE	77
SEX MATING	44
SEX PATTERNS	20
SEX THERAPY	45
SEX, SEXUALITY AND REPRODUCTION	7
SEX-LINKED INHERITANCE	40
SEXUAL ANATOMY	76
SEXUAL AND NON-SEXUAL REPRODUCTION	11
SEXUAL ATTRACTION	74
SEXUAL BEHAVIOUR	9
SEXUAL ENCOUNTERS	74
SEXUALITY: COMPLEMENTARY MATING TYPES	38
SEXUALLY TRANSMITTED DISEASES	61
SOCIAL AND CULTURAL ASPECTS	55
SOCIAL BEHAVIOUR IN ANIMALS REPRODUCTION	40
SYMBOLISM OF SEX AND THE LIFE CYCLE	7
SYSTEM, IN FROG, MOUSE, AND MAN ALIKE	18
TEACHING CHILDREN ABOUT SEX	77
TRANSSEXUALISM	17
UNISEXUALITY	43

BIBLIOGRAPHY - *Based on Books Published by Andreas Sofroniou*

1. Moral Philosophy, from Socrates to the 21st Aeon, ISBN: 978-1-4457-4618-0
2. Moral Philosophy, from Hippocrates to the 21st Aeon, ISBN: 978-1-84753-463-7
3. Therapeutic Philosophy For The Individual And The State, ISBN: 978-1-4092-7586-2
4. Philosophic Counselling for People and their Governments, ISBN: 978-1-4092-7400-1
5. Moral Philosophy, the Ethical Approach through the Ages, ISBN: 978-1-4092-7703-3
6. Moral Philosophy, ISBN: 978-1-4478-5037-3
7. Psychoanalysis, Poetry, ISBN: 978-1-4467-2741-6
8. Plato's Epistemology, ISBN: 978-1-4716-6584-4
9. Aristotle's Aetiology, ISBN: 978-1-4716-7861-5
10. Marxism, Socialism & Communism, ISBN: 978-1-4716-8236-0
11. Machiavelli's Politics & Relevant Philosophical Concepts, ISBN: 978-1-4716-8629-0
12. British Philosophers, 16th to 18th Century, ISBN: 978-1-4717-1072-8
13. Rousseau on Will and Morality, ISBN: 978-1-4717-1070-4
14. Hegel on Idealism, Knowledge & Reality, ISBN: 978-1-4717-0954-8
15. Philology, Concepts of European Literature, ISBN: 978-1-291-49148-7
16. Three Millennia of Hellenic Philology, ISBN: 978-1-291-49799-1
17. Cyprus, Permanent Deprivation of Freedom, ISBN: 978-1-291-50833-8
18. Sociology, Concepts of Group Behaviour, ISBN: 978-1-291-51888-7
19. Social Sciences, Concepts of Branches and Relationships ISBN: 978-1-291-52321-8
20. Concepts of Social Scientists and Great Thinkers, ISBN: 978-1-291-53786-4
    Medicine & Psychology
21. Medical Ethics through the Ages, ISBN: 978-1-4092- 7468-1
22. Medical Ethics, from Hippocrates to the 21st Century ISBN: 978-1-4457-1203-1
23. The Misinterpretation of Sigmund Freud, ISBN: 978-1-4467-1659-5
24. Jung's Psychotherapy: The Psychological & Mythological Methods, ISBN: 978-1-4477-4740-6
25. Freudian Analysis & Jungian Synthesis, ISBN: 978-1-4477-5996-6
26. Psychology from Conception to Senility, ISBN: 978-1-4092-7218-2
27. Psychotherapy, Concepts of Treatment, ISBN: 978-1-291-50178-0
28. Psychology, Concepts of Behaviour, ISBN: 978-1-291-47573-9
29. Psychology of Child Culture, ISBN: 978-1-4092-7619-7
30. Joyful Parenting, ISBN: 0 9527956 1 2
31. The Guide to a Joyful Parenting, ISBN: 0 952 7956 1 2
32. Philosophy for Human Behaviour, ISBN: 978-1-291-12707-2
33. **SEX, AN EXPLORATION OF SEXUALITY, EROS AND LOVE, ISBN: 978-1-291-56931-5**
34. I.T. Risk Management, ISBN: 978-1-4467-5653-9
35. Systems Engineering, ISBN: 978-1-4477-7553-9
36. Business Information Systems, Concepts and Examples, ISBN: 978-1-4092-7338-7
37. A Guide To Information Technology, ISBN: 978-1-4092-7608-1
38. Change Management In I.T., ISBN: 978-1-4092-7712-5
39. Front-End Design And Development For Systems Applications, ISBN: 978-1-4092-

7588-6
40. I.T RISK MANAGEMENT, ISBN: 978-1-4092-7488-9
41. THE SIMPLIFIED PROCEDURES FOR I.T. PROJECTS DEVELOPMENT, ISBN: 978-1-4092-7562-6
42. THE SIGMA METHODOLOGY FOR RISK MANAGEMENT IN SYSTEMS DEVELOPMENT, ISBN: 978-1-4092-7690-6
43. TRADING ON THE INTERNET IN THE YEAR 2000 AND BEYOND, ISBN: 978-1-4092- 7577
44. STRUCTURED SYSTEMS METHODOLOGY, ISBN: 978-1-4477-6610-0
45. INFORMATION TECHNOLOGY LOGICAL ANALYSIS, ISBN: 978-1-4717-1688-1
46. I.T. RISKS LOGICAL ANALYSIS, ISBN: 978-1-4717-1957-8
47. I.T. CHANGES LOGICAL ANALYSIS, ISBN: 978-1-4717-2288-2
48. LOGICAL ANALYSIS OF SYSTEMS, RISKS , CHANGES, ISBN: 978-1-4717-2294-3
49. COMPUTING, A PRÉCIS ON SYSTEMS, SOFTWARE AND HARDWARE, ISBN: 978-1-2910-5102-5
50. MANAGE THAT I.T. PROJECT, ISBN: 978-1-4717-5304-6
51. CHANGE MANAGEMENT, ISBN: 978-1-4457-6114-5
52. MANAGEMENT OF I.T. CHANGES, RISKS, WORKSHOPS, EPISTEMOLOGY, ISBN: 978-1-84753-147-6
53. THE MANAGEMENT OF COMMERCIAL COMPUTING, ISBN: 978-1-4092-7550-3
54. PROGRAMME MANAGEMENT WORKSHOP, ISBN: 978-1-4092-7583-1
55. THE PHILOSOPHICAL CONCEPTS OF MANAGEMENT THROUGH THE AGES, ISBN: 978-1-4092- 7554-1
56. THE MANAGEMENT OF PROJECTS, SYSTEMS, INTERNET, AND RISKS, ISBN: 978-1-4092-7464-3
57. HOW TO CONSTRUCT YOUR RESUMÊ, ISBN: 978-1-4092-7383-7
58. DEFINE THAT SYSTEM, ISBN: 978-1-291-15094-0
59. INFORMATION TECHNOLOGY WORKSHOP, ISBN: 978-1-291-16440-4
60. CHANGE MANAGEMENT IN SYSTEMS, ISBN: 978-1-4457-1099-0
61. SYSTEMS MANAGEMENT, ISBN: 978-1-4710-4907-1
62. THE TOWERING MISFEASANCE, ISBN: 978-1-4241-3652-0
63. DANCES IN THE MOUNTAINS – THE BEAUTY AND BRUTALITY, ISBN: 978-1-4092-7674-6
64. YUSUF'S ODYSSEY, ISBN: 978-1-291-33902-4
65. WILD AND FREE, ISBN: 978-1-4452-0747-6
66. HATCHED FREE, ISBN: 978-1-291-37668-5
67. THROUGH PRICKLY SHRUBS, ISBN: 978-1-4092-7439-1
68. BLOOMIN' SLUMS, ISBN: 978-1-291-37662-3
69. SPEEDBALL, ISBN: 978-1-4092-0521-0
70. SPIRALLING ADVERSARIES, ISBN: 978-1-291-35449-2
71. EXULTATION, ISBN: 978-1-4092-7483-4
72. FREAKY LANDS, ISBN: 978-1-4092-7603-6
73. LITTLE HUT BY THE SEA, ISBN: 978-1-4478-4066-4
74. THE SAME RIVER TWICE, ISBN: 978-1-4457-1576-6
75. THE CANE HILL EFFECT, ISBN: 978-1-4452-7636-6
76. WINDS OF CHANGE, ISBN: 978-1-4452-4036-7
77. A TOWN CALLED MORPHOU, ISBN: 978-1-4092-7611-1
78. EXPERIENCE MY BEFRIENDED IDEAL, ISBN: 978-1-4092-7463-6
79. MAN AND HIS MULE, ISBN: 978-1-291-27090-7

www.ingramcontent.com/pod-product-compliance
Lightning Source LLC
Chambersburg PA
CBHW072210170526
45158CB00002BA/526